Praise for

"A love letter and an elegy t[...]xist before the era of climate ch[...]rity of experience by a woman whose character was formed on the ski slopes in the Wasatch Mountains of Utah. Ayja Bounous is an emerging writer of conscience. She has a beautiful voice on the page and in place, this place 'shaped by snow.'"

—TERRY TEMPEST WILLIAMS,
author of *The Hour of Land*

"Those who love winter will love this book. But as we steadily erase the season that sets us free from friction, *Shaped by Snow* will appeal to anyone who has ever looked up and thrilled at the first flakes fat in the autumn sky."

—BILL MCKIBBEN, author of *The End of Nature*

"In this poignant and engaging love letter to snow, Ayja Bounous beautifully interlaces science, family history, and her anxiety around climate change. This book bears witness to the changing nature of Utah and the Wasatch Front, and is a call to all of us to pay more attention, to choose our actions with thought, and to live with love."

—SYLVIA TORTI, author of *Cages*

"The best people grow in open air, Walt Whitman told us, eat and sleep with the earth. Ayja Bounous is that person: raised on snow, seasoned on rivers, bound by conscience, called to action. This Utah oracle reminds us what's at stake, what we are fighting for."

—MARK SUNDEEN, author of *The Unsettlers*

"Young people take climate personally, and Ayja Bounous's Utah snow is personal terrain with a public dimension. Bounous mixes memory and desire to carve a bold line through the anthropocene slopes that surround her."

—JEFFREY MCCARTHY, Director of the Environmental Humanities Graduate Program, University of Utah

"Musings on powder, skiing, and the future of the Greatest Snow on Earth from a member of one of the Wasatch Range's royal families that is sure to appeal to Utah skiers."

—JIM STEENBURGH, author of *Secrets of the Greatest Snow on Earth*

"Bounous vulnerably shares her pragmatic yet emotional views on bringing children into the world we are crumbling, taking the reader through an introspective journey, connecting our passions to our past and our wishes to a very real future."

—BRODY LEVEN, professional skier

"There is no snow on earth like what funnels into Little Cottonwood Canyon and sustains the passionate powder hunters of Alta, Snowbird, and their adjacent peaks and bowls. *Shaped by Snow* is an intimate window into a heavenly place, by a writer whose family has rocked this cradle of skiing culture for three generations."

—NATHANIEL VINTON, author of *The Fall Line*

"A provocative read celebrating each and every snowflake yet leaving us with the question of environmental justice over the economics of the skiing industry and the future of life itself."

—BOBBI LYNN SMITH, Between the Covers Bookstore

SHAPED by SNOW

SHAPED by SNOW

Defending the Future of Winter

Ayja Bounous

TORREY HOUSE PRESS

SALT LAKE CITY • TORREY

First Torrey House Press Edition, November 2019
Copyright © 2019 by Ayja Bounous

Published by Torrey House Press
Salt Lake City, Utah
www.torreyhouse.org

MIX
Paper from
responsible sources
FSC
www.fsc.org   FSC® C011935

International Standard Book Number: 978-1-948814-10-2
E-book ISBN: 978-1-948814-11-9
Library of Congress Control Number: 2019932478

Cover photo by Sam Watson, "Winter in the Wasatch,"
    skier Sam Cohen in the Wasatch Mountains of Utah,
    www.samwatsonphotos.com
Cover design by Kathleen Metcalf
Interior design by Rachel Davis
Distributed to the trade by Consortium Book Sales and Distribution

*To little birds, wildflowers, barn owls, snowflakes, starlight,
and swallow-tail butterflies*

CONTENTS

❄ ❄ ❄

Introduction

In such anxious reflection as this, I crossed the bridge, em-
barrassed by my discourtesy in having appeared before you
without a New Year's present . . . Just then by a happy chance
water-vapour was condensed by the cold into snow, and specks
of down fell here and there upon my coat, all with six cor-
ners and feathered radii. 'Pon my word, here was something
smaller than any drop, yet with a pattern; here was the ideal
New Year's gift for the devotee of Nothing, the very thing for a
mathematician to give, who has Nothing and receives Noth-
ing, since it comes down from heaven and looks like a star.

—Johannes Kepler, *The Six-Cornered Snowflake*, 1611

In the winter of 1511, Brussels experienced such an intense
series of snowstorms that the entire city shut down for almost
six weeks. Instead of staying indoors, the residents took to the
streets. They began to pick up the white stars that fell from the
sky. They compressed the crystals, molding the snow into small,
spherical shapes that fit comfortably in their palms. They made
the snowballs larger and larger, stacking them on top of each
other, decorating them with stones and twigs, buttons and cloth.

The result? Hundreds of snowmen lining the streets of
Brussels.

If water is the key ingredient in the recipe for life, then snow is the zest that enhances the flavor. Snow forces life to be creative. Plants adapt to its seasonal pressures, figuring out ways to survive under what could be a few inches for a short amount of time, to a few feet for half a year. Some birds flee an area entirely when they sense snow in the air, while others drop their internal body temperature, shivering to stay warm. Reptiles and amphibians have yet to evolve the skills required to live actively in snow, sleeping the winter away instead. Many mammals hibernate as well, retreating into their dens to curl up in rich fur coats. Most, however, have figured out ways of continuing life during the winter.

In regions defined by snow, animal appendages become smaller while bodies become larger. Special kinds of fat develop during the food-rich summer months for energy storage during the winter. Pelts go from brown to gray to white, allowing camouflage. Some animals harvest the pelts of others, creating new fabrics to keep their bare skin warm. They build structures to protect themselves from the storms, and ignite fires for light and heat. They strap long, thin sticks to their feet and push themselves through the snow, the redistribution of weight preventing them from sinking too deep. This mode of transportation allows them to move as fast as the animals they hunt, helping them become more successful in the winter.

Snow sparks creativity. From forcing us to develop survival skills to inspiring us to create art, the six-pointed crystals have influenced humans for millennia. Snowfall requires us to work harder to stay warm and survive, but at times it releases us from our obligations and allows for leisure, as was the case in Brussels almost six hundred years ago.

Mathematician Johannes Kepler became intrigued with snowflakes in 1611. He was headed to a New Year's party when he noticed snowflakes landing on his jacket. He marveled at their delicate structures, mystified by why each had six points. Kepler hypothesized a number of theories, but never figured out the reason behind their starry shapes. His obsession earned his book of musings, officially titled *The Six-Cornered Snowflake*, the nickname "Kepler's Unsolved Problem."

The human-snow relationship is an unsolved problem. Some of us love it, *live for it*, while others are indifferent or hate it. Some have never and will never see it. Too much carbon in the atmosphere will cause snow to disappear from certain landscapes, yet our global climate relies on its cooling and reflective properties. The watersheds of the American West depend on it. Many of the cultures around the Wasatch Mountain Range in Utah developed because of its presence during the winter. I am a product of it, since my four grandparents and my parents all met and created lives together through snow.

I love snow. I love how it drifts outside my bedroom window and the way it covers surfaces, rounding out corners and smoothing the landscape. I stay up late into the night so I can witness how it glows after the sun has set. I love how the world feels smaller and domed after a storm, like being inside a crystal ball. As a child I spent hours writing my name in snowfields, my footprints creating designs in the fleeting substance. I'd ski every day during the winter if I could.

My relationship to snow is an unsolved problem. Climate change is threatening snow in the American West, causing moisture to fall as rain. When I ski, I participate in an industry reliant upon fossil fuels for operation and transportation. When I travel to the mountains, ride chairlifts, or spend time in resort buildings, I release carbon emissions into the atmosphere, contributing to climate change.

My family has helped develop the ski industry in Northern Utah for almost one hundred years. Skiing is the reason I live the way I live and have the relationships I have. It has heavily influenced the way I appreciate the world. Snow is one of the reasons I care about the climate. But by skiing, I contribute to snow's demise.

It's uncertain what the future of snow will be. Even if the world experiences horrific climate change, drowning islands and cities, and mass desertification, snow won't disappear completely. Water molecules will continue to crystallize and develop into snowflakes whenever the atmospheric conditions are right. But the way that we interact with snow will change. Organisms that have evolved around the constant, seasonal, or even intermittent presence of snow will be forced to adapt. Ecosystems reliant upon watersheds downstream of snow will struggle with drought. Communities, like mine, that have developed around the presence of snow will become scarce. We would no longer be able to line the streets with snowmen.

If climate change causes snow to stop falling, I won't just mourn snow. I will mourn the places that snow shapes: alpine ecosystems, glaciers and mountain ranges, watersheds and rivers, our climate, the beautiful ski community I grew up a part of. I will mourn that my grandchildren and great-grandchildren won't be able to catch snowflakes on their sleeves and marvel at their six points, or learn how to ski.

This book is my love song for snow; a way of sharing something that has shaped so much of this world and has made life on earth possible. It may also be my eulogy for snow; a way for me to remember it if it fades from my life, perhaps the only way for my great-grandchildren to experience it. This book is also a call to action: with my words, I fight for the survival of snow.

I. Blooming Season

Dreaming

On the longest day of the year, I wake up shivering. The brazen light of the solstice sun seems to nudge the nearby curtains aside as if they are nothing but a thin veil of smoke, drawing me out of a dream that I fight to hold onto. For a moment, my mind straddles two worlds at once. The brisk, frozen landscape I left in my sleep is much more familiar than the sweltering one I am waking in, with its pale yellow walls, gauzy curtains, and bright white cliffs in the distance.

My body tenses for a moment, my heart tapping in the hollow where throat meets breastbone. My chest rises and falls beneath the thin, white sheet. I struggle to remember why I am waking up here, and why I feel so cold despite the imposing summer heat.

The confusion lasts only a heartbeat as another pair of eyes opens inches away. Something in me registers their shape, the curve of their creases. Their cool, pale color holds the answer to both questions lingering in my groggy mind: why I am waking up in the desert, and why snow filled my sleep. With the recognition comes a swirling euphoria, a lightness that spreads from my chest to the tips of my fingers and toes, residual thrill and movement from the lingering dream.

Last night I dreamed I was skiing. I was moving effortlessly down the fall-line of a mountain, consumed by the cold smoke of powder. I was inhaling it. It was passing through the membrane of my skin and entering into my bloodstream. Water droplets froze and latched onto blood cells, creating icy creeks that turned into raging rivers in my legs, and the steady drip of capillary trickles in my fingertips. Frost crystallized on my eyelashes, my fingernails lengthened into icicles, my eyes froze over like the surface of a pond. My tendons became brittle like frozen bark, my lungs expanded with crystals. The cold connected muscle and sinew to chilled bones. It filled my womb and left frosted fingerprints in my hair.

Last night, I drove hours through a dark, winding canyon, trading mountains for desert. The man whose sharp, cyan eyes met mine this morning was the reason for my dangerous excursion, and, as I comprehend through the heat and haze of the moment, why I dreamed of snow.

Our fling started in the deep, dark months of winter. Keeping our developing intimacy to a minimum, our courtship consisted purely of skiing. We did enjoy the random, but intense, connection that our families shared with Snowbird Ski Resort in Northern Utah; our grandparents were involved in the development of the resort. And we created music together. During those winter months, we enjoyed each other's company in the mountains and in bed. When May arrived, Colin started his summer job as a raft guide in southern Utah and we had to part ways. I was uncertain of how our loosely formed relationship would hold over what would be close to four summer months of separation. There would only be a couple of chances to see each other, and each required one of us to take on the three-hour drive.

On this particular occasion, I had made the effort. He was in between rafting trips and couldn't leave the company's home base in Green River. The timing wasn't perfect, but it would be our only chance to be together for the next two months. So, in a spontaneous midnight decision, I risked the dark drive to the desert. That night, the summer heat retreated from my skin, the night grew longer and darker, and I felt frost creeping onto the window pane. That night, I dreamed of snow.

Despite the lazy days of the solstice, which draw on forever under the wide, yawning expanse of the sky, summer passes quickly, as summers often do. Upon returning to my home in Salt Lake City, the heat of the valley was nearly as stifling as the heat of the desert where I had said goodbye to Colin.

In Northern Utah, where the Salt Lake Valley butts up against the barrel chest of the Wasatch Mountain Range, summer can sometimes seem too long in the approach, everyone aching for the landscape to push through the hump of the mud season. Yet once it arrives, the heat that presses in around the scrub oak and sagebrush of the valley can be overwhelming, suffocating. It's in the mountains where the cooler air collects, settling on the granite rocks that give structure to creeks and streams, slithering between pillars of pines, wafting the rich, intoxicating scent of soil into the forest breezes. I crave these cool currents.

Still light-headed from those cold, dark moments with Colin, my skin aches for the touch of chilled fingertips and misted breath on my neck. So I leave my apartment in Salt Lake City and drive south to the mountains. I turn up Little Cottonwood Canyon and roll my car window down, twisting my fingers in the currents of the canyon breezes and playing music that reminds me of him.

For billions of years, water and fire have created the landscapes my family lives in; fire from the metamorphic processes beneath the earth and volcanic activity, water from shallow seas, glaciers, and erosion.

Little Cottonwood Canyon, home to Snowbird and Alta Ski Resorts, was created by a twelve-mile-long glacier. It extended from the topmost cirques of the canyon to the base, where it is believed to have butted up against Lake Bonneville, the Great Salt Lake's massive predecessor. My parents' house is just south of the mouth of the canyon, and every time I visit them I try to picture a glacier calving off into an ancient lake along the road I drive. Little Cottonwood Canyon has the distinct *U* shape common with glacial valleys. Running east to west, the canyon looks like an immense canal, connecting eleven-thousand-foot peaks to the valley floor. It's straight and open enough that someone standing in the valley can get a clear view of some of its highest peaks. In contrast, the glacier at the head of Big Cottonwood Canyon—the canyon north of Little and home to Brighton and Solitude Ski Resorts—extended only five miles. The upper sections of Big Cottonwood are wider, glacial valleys, but the lower section is winding and shaped like a *V*, since it was cut and eroded by a river. The two canyons are made up of similar rock formations and neighbor each other, but they each have their own distinct feel. Water shaped them differently.

The walls of Little Cottonwood are steep and severe. One of my father's favorite places to rock climb during the summer is the enormous granite slabs on the north side of the canyon. The granite rocks of the Wasatch were formed when intense heat began melting material beneath the surface of the earth forty million years ago. For twenty million years, eruptions hundreds of times larger than those of Mount St. Helens filled in the landscape above the earth's crust, while below it magma rose but cooled before it reached the surface, crystallizing and creating what are known as *igneous intrusions*. These intrusions are responsible for many of the igneous rocks in the Wasatch, like quartz monzonite and granite. Because of their light color, they stay cool to the touch even during the hottest times of the year. Whenever my father drives past these slabs, he slides his sunroof open so he can look up at the climbing routes. As his passenger, I hate when he takes his eyes off the road to look at mountains. But as the driver this June day, I lean my head slightly out of the window, trying to spot tiny climbers on the white-and-gray-speckled cliffs.

Hanging valleys, where a tributary glacier met the main glacier, bookmark the south side of the canyon. Waterfalls and creeks cascade from the lips of the hanging valleys until they meet Little Cottonwood Creek. The hiking trails leading up to those valleys are some of my favorite in the Wasatch. They curve around the canyon walls, weaving in and out of groves of aspen, keeping to the shadows of pines as they gain elevation. Many of them have creek crossings, where the temperature drops and the moisture in the air becomes tangible. Further up the valleys are cerulean lakes, and above those peaks, arêtes and cirques where glaciers cut jagged ridgelines into bedrock. These lakes and peaks are usually the destinations for hikers, but I don't care if I make it to them. I desire the coolness of the creeks.

I pull into a trailhead parking lot and step out of my car, breathing in the scent of rock and pine. On weekends, cars can overflow these small lots and line the canyon road, adding what appears to be a layer of shining, metallic scales along the black serpent of pavement. Today, however, I have the parking lot to myself. I begin gathering what I'll need for this hike—water bottle, notebook, protein bar. My hand hovers over a patterned rain jacket. Do I really have to bring it this time?

I glance at the sky above me. It's a blue so rich it's as though the mountains exist within a sapphire. I grab the crinkly thing and stuff it in my bag.

The Wasatch Mountain Range sits on the boundaries of three prominent geological features of the American West: the Rocky Mountains, the Great Basin, and the Colorado Plateau. The Wasatch crowns the Colorado Plateau, the high-elevation desert where I believe the heart of the American West resides. Shaped like an actual heart flipped on its side, the plateau is sliced open by the Colorado River, which moves massive amounts of matter that stains its water red through the ecosystem and out into the west, like an artery carrying blood through a body. Running north to south, the Wasatch is the most western range of the Rocky Mountains and the most eastern range in the basin and range pattern of the Great Basin, which spans from Utah to Eastern California, Southern Idaho to Northern Mexico. As the spine of the Uinta-Wasatch-Cache National Forest, the mountains are a sanctuary for mountain flora and fauna, residents of the Rocky Mountain ecosystem who depend upon the mineral-rich soil and constant water supply that the mountains offer in the Great Basin Desert.

The origin of the name *Wasatch* is disputed; some sources cite it as a Ute word for "mountain" or a "low place in high mountains," while others claim it comes from the Shoshone word for "blue heron." A rumor that the name comes from a Native American word for "frozen penis" circulates through the valley every few years. Sometimes I wonder why so many associate mountains and peaks with the intruding curve of a phallus. When I look at the Wasatch, I see the shape of a woman.

The word *cache* (of Wasatch-Cache) is more easily defined. A French word with Latin roots, a *cache* is a place to hide or to store things. It was brought to this region by French trappers, some of the first Europeans to venture as far west as the Wasatch, who used the mountains to store food, supplies, furs, and other tradable goods. Later, notorious robbers, like Butch Cassidy and the Sundance Kid, would use the mountains as hiding places. A *cache* might also refer to the natural stores of gold and silver that drew miners up into the mountains in the 1800s.

For many of the one million human residents of the Salt Lake Valley today, the mountains offer an escape from the twisting rivers of cement and boxy confines of suburbia, world-class backpacking, skiing, and rock climbing, and provide the valley with freshwater from snowmelt.

My family has lived in the Salt Lake Valley since the early 1900s. Though my mother's family hails from the Bay Area in Northern California, my father grew up on a fruit farm in Provo, Utah, about forty miles south of Salt Lake City. My sister and I were born in a hospital near the base of Little Cottonwood Canyon. Until we left for college, both choosing to head to the Bay Area rather than remain in Utah, we lived within two miles of the mouth of the canyon. These canyons and peaks, valleys and cirques are the places that shaped us, that we consider home.

Naming

Half of the bird was the color of a jewel, a bright cobalt that spread from its belly to the tip of its paddle-like tail. Its head and chest were a black onyx, with a few thin wisps of blue between its eyes. It had a crested horn of feathers on its head.

Mohawk, I thought as I stared into its beetle eyes, trying not to blink.

The bird ruffled its feathers once, twice, then took flight, leaving the pine branch swaying from its departure.

I left the cover of the trees and walked to the edge of the lake where my family was eating lunch.

"What kind of a bird was that, Baqui?" I asked. I don't remember how I gave my grandpa that nickname. He doesn't know either. It just emerged from my toddler lips one day, barely more than a babble. But it stuck. From that moment on he was Baqui.

My grandpa looked up from his sandwich.

"What did it look like?"

"Blue and black. It had a mohawk."

He chuckled. "Steller's jay."

I had seen plenty of those birds in my life, but never thought to ask its name until I was eight years old. It was a bird I associated with the mountains, one that could always be seen perched on the branches of pine trees.

We were hiking near Sundance Ski Resort with my father's side of the family. The broad shoulders of Mount Timpanogos, the second-tallest mountain in the Wasatch Range at 11,752 feet, towered above us, blocking the afternoon sun. The adults of the family, my parents, aunt, uncle, and grandparents, were lounging on the flat granite rocks near the shore of a small mountain lake. I had been playing hide-and-seek in the trees with my sister and cousins when I saw the bird. For whatever reason, I decided at that moment that I needed to know its name.

Potawatomi author Robin Wall Kimmerer says in her book *Braiding Sweetgrass*: "Names are the way we humans build relationship, not only with each other but with the living world." I had already developed a relationship with Steller's jays before I knew their name, recognizing the blue breast and mohawk whenever one landed in the scrub oak in our backyard. I knew its call, a throaty screech that doesn't match its physical beauty. And I could tell the difference between a Steller's jay and a scrub jay, a close relative with feathers a lighter shade of blue and a soft gray belly. But naming it did give me a certain awareness of the bird, a new kind of connection. Now whenever I saw it I'd say its name.

Steller's jay.

A few years later we were hiking that same trail with those same cousins. I was carrying a large, gauzy net with a long, wooden handle. There was a mason jar in my backpack with hardened paste at the bottom, soaked with nail-polish remover—my "kill jar." I was the straggler that day, falling behind the rest of my family as we switchbacked up the side of a steep slope.

"Ayj, you're holding up the group behind you!" my father called from a few switchbacks above.

He was right. The group I was stalling was eyeing me hesitantly, unsure if they should try to pass or not. Or maybe they were uncertain what to think of the preteen waving a net as tall as she was through the grass along the side of the trail.

My new obsession was bugs. I had been given a biology assignment that summer to catch insects, kill them, and categorize them. I didn't like the killing part, but I loved learning the scientific names. Beetles became *Coleoptera*, dragonflies *Odonata*, bees *Hymenoptera*, grasshoppers *Orthoptera*. I took the net with me wherever I went.

I jogged up a few switchbacks to catch my family. My cousins were leading the charge and singing as they hiked, their voices strong despite the elevation. Panting, I fell in step behind my grandpa, the soles of my shoes padding softly on the dense mountain dirt.

Much of the western United States is almost a mile above sea level. We were hiking at an elevation of about ten thousand feet that day. Over ten million years ago the western US began to rise, expanding as it did so. Like a cake rising, the crust of the earth cracked as the expansion occurred, creating faults throughout the West and forming the basins and ranges of the Basin and Range Province. The most active of these faults is the Wasatch Fault, responsible for the drastic eight-thousand-foot elevation difference between the peaks and the valley floors. The intensity of the Wasatch Fault is such that, if not for the constant erosion by the elements, the peaks in the Wasatch could be forty thousand feet in elevation rather than the twelve thousand they are today.

Wildflowers lined our path, delicate and colorful. There were many I couldn't name by sight, but I did recognize the yellow black-eyed Susans and glacier lilies, white-and-pink columbine, scarlet-and-tangerine Indian paintbrush, and my favorite, deep-purple lupine. They were blooming a little earlier this summer than they normally would. Wildflowers in the Wasatch typically bloom mid-July through August, but Northern Utah had experienced a low snow year the prior winter and the snow in the mountains melted quickly. Once the soil is exposed and saturated with moisture from the snowpack, seeds that have lain dormant in the ground since the fall are able to germinate. When the snowpack is thicker, it lasts further into the summer, pushing the wildflower season back.

After one particularly heavy winter, the wildflowers didn't bloom until September. The snowpack lasted so long into the summer that my parents skied on the Fourth of July—Snowbird's closing day. My grandparents went on a hike on my grandma's birthday, September 18th. Mid-September in the mountains is usually well past wildflower season, when the leaves start changing colors. But that year mid-September was prime blooming season. In a picture taken on their hike, my grandparents stand in a thick field of flowers that reach their knees, the persimmons and indigos of Indian paintbrush and lupine so dense you can't see their shoes. "Wildflowers in Mineral, September 18, 2011. Fast Max's 86th birthday" is written in sharpie at the bottom of the photo.

Fast Max is my grandma Maxine's nickname; she may have never broken the 4'11" threshold in her life, but she was known for her speed on the mountain, both while hiking and skiing. I might call him Baqui, but my grandfather's proper name is Junior Bounous. However, Junior wasn't his legal name until he was almost thirty years old. The youngest of a large Italian family, no one came up with a name at the time of his birth. His birth certificate officially read "Boy Bounous." Baqui didn't know that Junior was a nickname until his wedding, when he pulled out his birth certificate. Now, the name Junior Bounous is woven with folklore of Wasatch snow and deep powder skiing.

"Baqui, what's your favorite wildflower?" I asked after regaining my breath.

"Favorite? Oh, I don't know if I have a favorite," he said in his slow voice.

My preteen self, in the habit of ranking favorites for everything, from ballpoint pens to polo shirts, didn't accept that for an answer.

"Well, which one do you like more than the others?"

He gave a chortle, glancing at the wildflowers near his feet.

"I guess I do love elephant's head."

"What does that look like?"

"Mmmm, they're pink. Pinkish purple. Little flowers that look like elephant heads."

I didn't try to get any more out of him.

We made it to our destination, an alpine lake called Emerald Lake. Mount Timpanogos loomed to the west, striated cliffs and ridges imposing and magnificent. Most of the rock in this area is limestone, created during an era of heavy sedimentation. Many shallow bodies of water sprawled over the American West between three hundred and five hundred million years ago. Sediment, in some areas three miles thick, was deposited at the bottom of these seas, creating limestone, shale, and siltstone. During two ice ages that occurred thirty thousand and twelve thousand years ago, enough snow fell on this landscape that glaciers formed. They carved out the canyons of the Wasatch. Emerald Lake is one of the remaining features left by the glacier that made this mountain.

Above the lake is another feature left behind by an ice age. A mass of ice and rock slowly creeps down the side of the mountain: a rock glacier where there may have once been a proper glacier. Up until the last century, when a warming climate and more dust in the atmosphere began affecting snow in the Wasatch, the rock glacier had a permanent snowfield that lasted through the summer. Skiers would hike 3,500 vertical feet to ski it. My grandparents were two of those skiers.

My father brought my mother up to this snowfield on one of their first dates. It was early in the spring so they were still hiking on snowpack, but the snow had melted enough that there was a gap between the rock glacier and the mainland. According to my mother it was about five feet wide, and a fifteen-foot drop to a raging creek beneath. In her words, "It wasn't huge, but it was one of those situations where if you missed your step it'd be the last thing you'd ever do." My dad picked up the two dogs they were hiking with and threw them across the gap. They tumbled on the snow and stood up, apparently unfazed. Then my dad jumped, without checking to make sure my mother was comfortable following. My father was a professional racer on the US Ski Team at the time and in prime athletic condition. He had an adventurous soul, a way of getting people into situations that might make them feel uncomfortable, but also a way of getting them out. My mother was taken aback, but knew if she thought too hard about it she'd freeze. So she jumped.

We were finishing our lunch at Emerald Lake when my mother suddenly grabbed my net off the ground. She leapt into a field of wildflowers, sweeping the net in front of her. She had a triumphant expression on her face when she returned.

Trapped inside the net was a beautiful black and yellow butterfly. *Lepidoptera*.

It was much larger than any of the insects I had caught so far. I was a little shocked; I had never really thought my mom to be the bug-collecting type. She smiled when I said as much.

"I wasn't really into bugs. But I liked picking up bones on the side of the road."

I was torn. The creature was so beautiful, with long, delicate wings. I didn't want to kill it. But my biology teacher would probably give me extra points.

"Baqui, what kind of a butterfly is this?"

My grandpa ambled over, placing his hands on knobby knees as he bent over the net.

"Swallowtail. See these tips on its bottom wings? They look like the forked tail of a swallow."

I placed it in the kill jar as gently as I could. It was too big even for the jar; its wings flapped against the glass. I slid it in my backpack and tried not to think too hard about my decision on the hike down.

I shook its limp body out of the jar once we got home. It was gorgeous, its antennae delicate and curved. Veins of black fanned across thicker spots of yellow like spiderwebs. At the bottom of the wings were little grayish-blue smudges, and two small red dots which I guessed were meant to look like false eyes to distract a would-be predator. I stroked its body with a fingertip. It was so soft. I placed it on the foam board my teacher had given me, using pins to spread its wings to their full extent, and measured it to be larger than my palm. I left it there overnight, letting the body dry out in that position before I'd be able to add it to my growing collection.

I woke up the next morning and went to move it. The butterfly was alive, flapping futilely on the foam board. It had holes in its wings where the pins were still stuck through them, and another pin stuck in its body. My insides burned as I watched its movements. Tears slid down my cheeks as I placed the kill jar over it, making sure it was sealed this time.

Years later, at my high school graduation ceremony, my principal introduced me in front of an entire concert hall.

"Ayja Bounous had the best bug collection."

Emitting

When I was a child and my father had retired from being a professional racer to become the director of the Snowbird Ski Team, one of his ski coaches was nearly killed by an avalanche while walking out of the old coaching shack at Snowbird. The shack was at the top of a beginner run called Chickadee, built right at the base of a peak named Mount Superior. As Mark left the protection of the shack he heard a deafening rumble to his right. A churning mass of white had dissolved the face of Mount Superior and was tumbling toward him at impossible speeds.

He dove back into the shack rather than jumping in his car. His instinct saved his life. The avalanche took his car over the lip of the road and down onto Chickadee. The small wooden structure, reinforced with large logs, remained intact.

During the summer, the peak's iconic appearance barely hints at the avalanche risk it poses during a different time of year. Superior is composed of cottonwood tillite, glacial till that's been pressurized into solid rock, and tintic quartzite, metamorphic granite. The combination of the two creates waves of rust and cream that colors many of the peaks in the Wasatch. Around one hundred million years ago, the North American continental plate, which had previously been traveling east, switched directions, colliding with the denser plate that the Pacific Ocean is on. One plate sunk beneath, nudging the other skyward. The landmass that we consider the western United States was lifted up, one reason why much of the western US is almost a mile above sea level. The collision resulted in an era of mountain making. The iconic waves on the face of Mount Superior, most noticeable in the rusty, tillite formation, are what remains of the ancient mountain range. Since the peak is covered by snow in the winter, this pattern is most prominent during the summer, and adds dramatic movement to the mountain's stony appearance.

To the east of these waves are gray-and-white-striped cliff bands of limestone and marble, which contrast the neighboring creams and dusty reds of Superior. The feature is a popular climbing destination called Hellgate. The first avalanche I saw as a child was one that careened over those cliffs.

I was once hiking with a friend along a trail at Snowbird, directly across the canyon from Mount Superior. Looking at Superior's auburn-streaked face, he remarked that the mountain was much more beautiful during the winter. For reasons unclear to me in that moment, I was offended by his comment. I couldn't comprehend how he could compare the two. I grew up as close to the mountains during the summer as I did during the winter, and for me to choose between them would be like a mother trying to pick a favorite child.

I once asked Baqui if he had a favorite season. He responded: "Yes. It's called the four seasons."

Superior is a popular peak to summit during all seasons. In the winter, skiers and boarders will skin up it to ski down, making it some of the most sought-after backcountry terrain up Little Cottonwood Canyon. During the summer, many hike to its peak along a more gradual ridgeline, or scramble up the south ridge. The south ridge is exposed and a bit treacherous. A rope and climbing gear aren't necessary to climb it, but a level of athleticism and sure-footedness is. When my parents climbed it, my mother asked my father to rope her up during a certain section, just in case she were to lose her footing. He attached the rope to her and she climbed the section confidently, only to realize he hadn't actually been holding onto the other side, letting the rope drag behind her. My mother said that the summit of Superior was gorgeous, but made even more beautiful by the thrill and exposure of the scramble. She said they were hiking the edge of fear and excitement.

Of the places and peaks I've been to in the Wasatch, I haven't made it up Superior yet. I tried one summer, but there was so much smoke in the air that I decided against it. A few of my friends hiked during that time, and said their lungs and throats burned with the effort and they felt sick afterward.

The mountains aren't quite the escape from the valley heat that they used to be during summer months. Wildfires have impacted the quality of living in the Salt Lake Valley in recent years. Some of the smoke pollution is from fires in Utah, while some smoke blows in from as far away as California and Oregon. Fires are a common occurrence in the West during late summer and fall, and they've been getting worse. The summer of 2018 was the driest year ever recorded in Utah and one of the worst summers for air quality from the resulting smoke. And fires are predicted to get more violent in the American West as climate change continues to affect our atmosphere.

Climate change is a disputed topic but it shouldn't be. I'd wager that if not for the money in the fossil fuel industry, climate change wouldn't be a debate. There is little doubt in the scientific community that climate change is happening and that humans are contributing—and that the consequences will be catastrophic.

When sun rays enter the earth's atmosphere, some of the light is reflected by lighter-colored matter, like snow and clouds, but most is absorbed by the oceans, soil, vegetation, and rocks, as well as things like parking lots and infrastructure. The energy from this light is then radiated back out to the atmosphere as infrared heat. Particles in the atmosphere, like carbon dioxide and water molecules, absorb this energy and radiate it in all directions and back toward earth. Known as the *greenhouse effect*, this phenomenon is what makes our planet warm enough to live on, and is responsible for the abundance of life on earth.

The particles in the air that contribute to this effect are called *greenhouse gases*. Some particles in the atmosphere react either physically or chemically to the warming, while other particles don't react. Particles that do react don't control our climate but rather are controlled *by* it, and contribute to a feedback system. Particles that don't react are called *forcing gases*, and are the most important regarding climate change.

Water vapor is the most common feedback particle. Water molecules end up in the atmosphere by evaporation from bodies of water and plant transpiration. The warmer the temperature of the air, the more water vapor the air is able to hold. As evaporated water rises into the atmosphere, it can cool and condense from gas back into liquid. This is how clouds form in the atmosphere; excess evaporated water in the form of gas condenses to become clouds and eventually rain or snow.

Greenhouse gases that "force" climate change, like carbon dioxide, are ones that don't react to temperature changes, like water does. When heat is radiated into the atmosphere, carbon dioxide doesn't change form with the energy, but instead blocks it from leaving the earth's atmosphere and radiates the heat back to the surface of the planet, causing more warming. And as planet-wide warming continues it leads to more evaporation, and more water vapor in the atmosphere, which in turn causes more warming. Water and carbon dioxide play off each other to enhance climate change, but if not for carbon emitted from humans, this effect wouldn't be so great.

It doesn't take much carbon dioxide to cause this. Carbon dioxide is measured in the atmosphere by parts per million (ppm). Pre-industrial-era carbon concentration was around 280 ppm. This means that for every one million molecules of other gases in the atmosphere, there were 280 molecules of carbon dioxide. This is the level that the current life-forms on our planet evolved to survive in. By the environmental group 350.org's estimate, 350 ppm is a reasonable and healthy goal for our planet—hence the organization's name. Currently, we're over 400 ppm. It's likely that we've passed 400 ppm for the last time in modern times, and we're releasing approximately 2 ppm into the atmosphere each year. Once we pass 450 ppm, scientists believe, we will not be able to mitigate the earth's warming.

Large swaths of materials on the earth sequester carbon dioxide, storing it for long periods of time and preventing it from entering the atmosphere. When they absorb more carbon than they release, we call them carbon sinks. Dense forests are a type of carbon sink; trees and plants take in carbon dioxide during photosynthesis and convert it into starch. Oceans are another. Dissolved carbon dioxide in the oceans gets used by organisms like phytoplankton for similar purposes. An important carbon sink is the permafrost, soil in places like the Arctic Circle that's been frozen for at least two years, though much of it's been frozen for tens or hundreds of thousands of years. Frozen organic matter in these soils contains large amounts of both carbon dioxide and methane.

As the planet warms, these carbon sinks are becoming carbon emitters. When we cut down forests, we not only remove organisms that pull carbon dioxide from the air, but we usually burn them or convert them into other materials, releasing carbon emissions. The ocean is one of the planet's most important carbon sinks, absorbing approximately 30 percent of carbon emitted by humans. But as the temperature of the ocean warms, its ability to dissolve carbon diminishes. And when the permafrost melts with the heating of the poles, which are some of the fastest-warming places on earth, carbon in the form of both carbon dioxide and methane could be released in a massive "burp." Methane is one of the most potent greenhouse gases, trapping heat about twenty times more effectively than carbon dioxide does. There is almost twice as much carbon stored in the permafrost than there is in the atmosphere today, so the melting permafrost will have very alarming consequences for climate change.

Along with the poles, mountains are already feeling the effects of climate change more than places at lower elevations. Elevations over thirteen thousand feet are heating up to 75 percent faster than elevations under six thousand. Winters are expected to warm four to ten degrees by the end of the twenty-first century. Alpine ecosystems might be hardy when it comes to surviving frosts, blizzards, and thick snowpack, but they're delicate when it comes to climate change.

The cycle of life in the mountains is timed with the snowpack. When a snowpack melts early, it exposes the soil to the elements. This allows vegetation to start growing sooner, but it also means that the new growth becomes susceptible to frosts, which can keep occurring in the mountains through June. Early snowmelt also means that alpine soil dries out quicker during the summer months, which affects pollinators, insects, and animals reliant on plants for sustenance, and secondary consumers who rely on those insects and animals for food. When high alpine meadows experience stress from warming temperatures and drier environments, more tolerant plants establish themselves during those vulnerable times, disrupting mountain ecosystems.

The snowpack in the mountains provides up to 75 percent of the freshwater in the West. We might not be able to see large-scale carbon emissions in the atmosphere, but we can certainly feel the effects of it here in Utah. Desert ecosystems, mountain ecosystems, and agricultural land, all reliant on snowfall, are under threat from climate change. As are our forests, which are becoming more susceptible to wildfires from less moisture. And with forest fires comes air pollution, smoke, and hazy skies. Bad for hiking, terrible for our health, destructive to our land. Deteriorating snowpack from climate change is one of the biggest threats to life in the West.

Building

"Ready?"

I shouldered my backpack and nodded to my mother.

"Ready."

If there was ever a mountain goat in my family, it would be my mother. Slight in build, shorter and smaller than me, her size is deceiving when it comes to hiking. As I fumbled to adjust my bag, she propelled herself down the trail.

Our path curved around the right side of the lake before switchbacking up a steep slope. Just before the trail cut upward I glanced left, to the far side of the lake, to see if I could get a final glimpse of my father before we began our climb. But he must have already found a trail into the dense vegetation that lined the shoreline, for he was nowhere in sight.

The sky above us was the shade of blue I always associated with the Wasatch, a brilliant azure framed by rust-colored peaks and emerald pines. But gray clouds threatened its edges. Their foreboding color didn't just imply rain, but thunder. And hail.

We could sense, rather than hear, the rumbles coming from the other side of the ridgeline, as though the thunder was coming from deep within the earth rather than the sky. The sound wasn't loud, but its muted timbre made it seem even more ominous.

We had started our hike from a friend's home in an area called Grizzly Gulch, located at the top of Little Cottonwood Canyon. It was a gorgeous summer morning in the mountains, temperatures in the mid-eighties, no storms in the forecast. My grandparents joined us until we reached the ridge that separates Big and Little Cottonwood Canyons. My parents and I continued down to a series of lakes, where we stopped to eat lunch.

My father got a call from Baqui while we ate, warning us that he and my grandma had just been caught in a hail storm. It had moved in from the west, and lasted about two minutes. If I were just visiting the Wasatch I may have thought he was joking. But having grown up in its shadow, I knew how unpredictable the weather patterns could be. Summer afternoons are wildly deceiving, when thunderstorms can build and break on the crowns of the Wasatch while the valley below remains sunny.

Uncertain if we should chance hiking up to the exposed ridgeline or remain by the cover of the trees, we finished our lunch quickly, listening to the high *trillllll* of a spotted towhee. There was still no sign of any storm. The nearby peak cast its reflection into the calm lake at our feet, a prism frozen in glass. According to my grandparents, there was a hail storm just on the other side of its ridgeline.

Then my grandfather called again, reporting that another brief storm had hit, this time with lightning.

"Where are these storms?" I asked as my father hung up.

But even as I spoke we heard the barest growl of thunder, without any hint at which direction it came from. We gathered our things. My dad thought he remembered a path that could get us back faster, but didn't quite know where it started or the state it was in. My mother and I shared skeptical glances; my dad was notorious for convincing people to follow him on a "short-cut" and then winding up on a cliff band or stuck in a marsh, a scenario we lovingly call "Bounousabuse." When he admitted there might be some bushwhacking involved to find the trail, my mother and I decided not to chance it.

Wishing us good luck a few minutes later, he took off down the left side of the lake while we headed right, starting our climb as the barest hints of the storm began peering over the ridgeline.

When my mother is determined, or frightened, she can set a pace that even men with much longer legs than her have trouble keep-ing up with. Panting, I kept my head down, focusing on the trail, wiping sweat from my brow every few minutes. The sun con-tinued to press down on us, and each time I looked up the gray clouds seemed to be perched right on top of the ridgeline, not yet spilling into the basin. But the thunder was growing louder.

We were approaching a shoulder, a nook in one of the ridges that led to the peak above us, when my mother turned to say something.

"Looks like—oh." Her words disappeared into the breeze as she looked behind me in shock. A wall of moisture smeared the basin we had climbed out of. While we focused on the storm in front of us, the one my grandparents reported, another had moved in behind us. The sun had somehow remained in a patch of blue sky, just above us, giving us a false sense of comfort while clouds encircled the basin. Spidery fingers of lightning flickered through the gray streaks of rain across the valley.

Leaving the trail, we beelined toward a little cliff band with some tree cover, ducking into a grove right as the sky broke over us. It hit so quickly we didn't have time to take out our rain jackets, and as we bent over our bags to grab them out, hail the size of pomegranate seeds began stinging the backs of our necks. Thunder shook the ridgeline menacingly. Lightning silhouetted the few trees that stood around us.

"Do you think this is a safe spot?" my mother asked over the sounds of the storm.

"I don't know," I responded. "Do you?"

The rocks we stood by were not so much a cliff band as they were large, straight boulders, freestanding in the shoulder of the mountain. There were a few pines around us, probably six or seven, and they weren't providing much protection from the hail. We were surrounded by the tallest things in the vicinity.

Discussing our options, we decided that the grove was still safer than retreating back down the exposed path we had just hiked up, or continuing on the trail across the face of a mountain. I attempted to remember everything I had ever heard about lightning, though I knew much of it was myth. Getting rid of our hiking poles seemed like it wouldn't hurt, so we tossed them into some bushes thirty feet away, then hunkered down beneath the trees again, keeping our feet close together.

I once asked my father if he was scared of anything. He always seemed so fearless, jumping off of cliffs on skis, rock climbing, windsurfing chopped-up ocean waves.

"I'm scared of things you can't control," he said. "Things like lightning, or avalanches. You can do your best to stay out of situations that will get you into them, but they can be unpredictable. That's what scares me."

My family has a history with lightning. When my dad raced for the US Ski Team in the 1980s, one of their races in New Zealand was interrupted by a lightning storm. My father was at the top of the course when a buzzing noise started; he fled when he noticed a man's hair standing straight up. Lightning struck my grandparents' car once while they were driving through southern Utah. My parents were near the summit of Kings Peak in eastern Utah when my mom's braids started curling up at the ends, and my dad's hair rose from his scalp. They had also been caught in a storm on the side of a cliff while rock climbing a multi-pitch, anchored to the wall by metal pieces and no quick way to get down. And around the time I was born, a family friend's son was killed by lightning while camping in southern Utah. The storm had been miles away from his campsite.

I could almost see my mother replaying all of these events in her mind as we stood in the trees. Hail bounced around our feet. The balls grew in size as the minutes passed, smacking our shoulders and jacket hoods relentlessly. The thunder and lightning were now simultaneous. My legs, bare beneath my shorts, quickly went numb, and within minutes there was almost an inch of slush around our feet. My light running shoes soaked through immediately. This was not the two-minute storm my grandfather had reported. I leaned up against the granite rock, then, noticing the black-and-white veins in it, pushed myself away, wondering if certain types of minerals and rocks attracted lightning.

Another *flash-boom*, so blinding and awful that I was tempted to fall on my knees and crawl away—where? There was no escaping this.

"Look!" my mom exclaimed, pointing up the ridgeline.

Above us, about a hundred feet away, a tree was softly smoking through the haze of hail.

Forty-five minutes had passed since we took shelter in the shoulder, and still the storm showed no sign of retreating. Each time it seemed as though the storm was getting further away, a new round of lightning would roll in. Too scared to touch our phones, we had not tried to contact my dad to see if he had managed to get to safety. We stood in four inches of hail slush, and I was beginning to feel feverish from the cold. The only purpose our rain jackets really served anymore was lessening the impact of the hail on our skin.

One . . . two . . . boom!

Now desperate to get off the mountain, my mother and I had started silently counting between flashes and thunder. It was rare to make it past two.

Flash!

I watched my mother's mouth form an O, but the thunder came before she could finish mouthing "one."

After the next flash we counted *one . . . two . . . three . . .* and my mother's eyes widened. The trail ahead did not continue on the ridgeline but it still looked dangerous, cutting across the face about a hundred yards beneath the summit. We'd be the tallest things for almost a half a mile. But the situation had become dire enough that we needed to move, and we still had another three miles of hiking to get back to our friends' house. Those miles, at least, were beneath the cover of thick forests.

"Ready?" she asked.

"Ready."

We dashed out of our hiding spot, abandoning our poles to the bushes we had thrown them in earlier. We were almost halfway across the hill when the world broke apart in light and sound. *Flash-Boom!* We stopped running, our instincts dropping us into a crouch. There was nothing around us, no kind of shelter. Even crouching, we were the tallest things in the vicinity. Each slow second passed with fear that the next second might be our last. I suddenly and fully comprehended my parents' fear of lightning.

My mother looked up at me, terror in her eyes, and said, "What do we do?"

My mother had always fiercely protected me, almost to a fault. If I came home from school upset from something someone had said, she'd be on the phone within minutes to call up the parents of whichever kid made me cry. When I was so shy I couldn't talk to waiters, she'd order my ketchup for me.

And now she was turning to me to make the decision. Our eyes held each other for a long moment. I briefly wished my father was there to choose for us. Despite his daring that gets us into uncomfortable situations, he maintains level-headedness while our composures dissolve.

Flash-Boom! Flash-Boom! Flash-Boom!

I looked both ways. We were definitely closer to the shoulder we had just left. We could run back and continue hiding, or we could risk the longer route to shelter. But there was no way we were staying there, out in the open, waiting for someone else to make the decision.

"Let's run to the woods," I said, pointing to make sure she understood. "Ready?"

She nodded, her eyes unblinking in her pale face.

"Ready."

It felt as though hours passed, but we were beneath the cover of the trees within two minutes. As the adrenaline faded we began comprehending how cold we were. The declining path was flooded with hail that had partially melted into slush, and we slipped and fell all three miles to the house.

My grandparents and father were waiting anxiously when we arrived. My dad had found his trail, more protected than the one we were on, and made it back two hours before us. My mother and I were both so weak it was all we could do to strip out of our wet clothes and into the dry ones they gave us. I felt flushed and nauseous, and a headache was attacking my head and neck. My limbs ached. It took hours before we felt good enough to drive home.

I had experienced unexpected Wasatch storms that had sent me running for cover many times in my life, but none as ferocious as that.

Breathing

There is a place at Snowbird named after my grandpa that's home to some of the most beautiful wildflowers in the Wasatch. It's a small, cupped hand, the upper cirque of a glacier long gone, called Junior's Powder Paradise. Powder Paradise, its common name, is the upper section of Mineral Basin, the same place where my grandparents posed for a picture among the wildflowers. A ridgeline leading to Twin Peaks encircles almost 180 degrees of Powder Paradise. In the summer it contains a lazy stream which, due to high arsenic levels, flashes bright copper back at the sun. Dense, green moss encroaches upon the shores of the stream. The water gathers at the lip of the cup, becoming a small pool before ducking under the basin's rim, disappearing into the mountain, and reemerging as small, babbling waterfalls on the other side.

It's named after my grandpa because of the role he played in helping develop Snowbird. He worked at a number of resorts during his ski career, beginning with Timp Haven, now called Sundance, then Alta, Sugar Bowl, and finally Snowbird. When Snowbird was being developed, Baqui helped design the runs, deciding how skiers would interact with the landscape as they traveled down the mountain. He started the ski school as well, and held the title of Director of Skiing until he retired at ninety years old.

One of my grandpa's favorite trails in the world starts at the top of Hidden Peak and descends into Powder Paradise. He's summited peaks and backpacked through deserts, but this beautiful yet simple gathering of copper and green, accessed only during the summer through the gates of paintbrush and lupine, holds his heart within its moon-shaped curve. I couldn't tell you why exactly; that's between him and the mountain.

It's not just during wildflower season that Baqui loves visiting this place. During the winter, when white has paved copper into hibernation, he visits this bowl even more frequently. It's named Powder Paradise for a reason.

After one particular run, when life and location came together perfectly, he connected, however momentarily, with this place. He called it a "perfect run," but he didn't mean that his skiing technique was perfect or the snow was the best he's ever experienced. He couldn't describe what happened that day, but it seemed to me as though his body had merged with the snow and the basin for a few moments before retreating back into itself. It was a deep, profound, and perhaps slightly frightening moment for him, one that twisted into his body and mind.

My grandpa couldn't find a word to describe how he felt. I'd wager that most humans during their lifetimes will have experiences like his, when we connect with another human, animal, or place. Over the years, some have attempted to assign words to these fleeting encounters.

While looking into the eyes of a bighorn sheep, author Ellen Meloy says, "There is in that animal eye something both alien and familiar. There is in me, as in all human beings, a glimpse of the interior, from which everything about our minds has come. The crossing holds all the power and purity of first wonder, before habit and reason dilute it. The glimpse is fleeting. Quickly, I am left in darkness again, with no idea whatsoever how to go back."

Once, while I was gardening in my backyard, I stopped to take a break from bending over the garden bed, sitting down and digging my hands into the soil between two rose bushes. I sat there, still and quiet, for a minute or so. Two California quails, one male, one female, came out from behind one of the rose bushes, pecking at the ground as they did so. They startled me, but I managed to keep still, holding my breath. The male noticed me first. He straightened abruptly, gray-blue chest puffing out, the dangling, black hook on his head shaking. The female straightened next, her light brown plumage soft and feminine next to her mate's bolder stripes of white. They both glared at me sideways, their heads cocked at stiff angles. I imagined I could sense alarm and hesitation in their stare. But I thought I caught a sliver of something else in their eyes, just out of reach, a type of unattainable knowledge. No sooner had the thought crossed my mind when they both flew off the ground, their short, fat wings beating so violently in the small space that I recoiled, my shirt snagging on the rose bush behind me. I was left in the stillness and silence after their departure, my heart beating viciously in my chest.

Philosopher Martin Buber separates these moments of truth from the experiences in our day-to-day lives by defining them as *encounters*. He says that when an encounter occurs, it can only last for a moment, because the minute we realize that we're having a moment of enlightenment the encounter fades, returning to an experience. For Buber, a moment of encounter means a moment of knowing what God truly is: love.

Back in the 1700s, the word *sublime* was used by scholars like Edmund Burke and Immanuel Kant to describe these moments, specifically when they occurred in the wilderness and simultaneously included terror and pleasure. While climbing Mount Katahdin in Maine, Henry David Thoreau had a moment of disorientation while connecting with the wildness of the mountain, and exclaimed that he felt *contact!* with the rawness of the landscape.

Ellen Meloy would call the enlightenment a *glimpse*, Martin Buber would call it an *encounter*, while some scholars would call it a moment of *sublime*. Thoreau might call it *contact!* Adding my own voice to the choir, I would call it *birth*.

Whatever it was, Baqui told me that a single tear emerged from his olive eyes in that moment.

I have a theory about souls. I believe that every living thing has a soul, and our souls create the fabric of life, which is sewn into the landscapes of the earth through vibrations, or energy. The evolution of life was the marriage of the living fabric to the material landscape, the product creating a soul. Living bodies, which are composed of the vibrating matter of the landscapes they originate in, borrow patches of souls from this energy source, combining physical material with life, and once the body is no longer living, the soul returns to the landscape it originated from, a "soul's place." Along the course of life, a soul picks up vibrations from other places. This energy, while held snugly over the world, is constantly shifting and moving due to the movement of souls. Nothing is separate. Our bodies are composed of physical matter that we borrow from the earth.

And when we have those *encounters*, those *glimpses*, those moments of *birth*, we are collecting energy, our souls merging with the landscape and bringing new material into our existence.

Powder Paradise is my grandpa's soul's place: the womb from which he originated.

Growing

Glass bottles clinked against each other as I shrugged off the terry robe. Hanging it on one of the wooden pegs lining the wall, I pulled the beer bottles, already slightly wet with moisture, out of the large pockets of the robe. One of my cousins from my mother's side of the family, Reyna, her three-year-old daughter, Kyler, and my mother were already in the pool.

It was a gorgeous, late-summer's day at Snowbird. My cousin had flown in from Seattle with Kyler earlier that week and we were staying in a set of rooms at the Lodge, one of the hotel accommodations up Little Cottonwood Canyon. They were here for the Oktoberfest Snowbird hosts every August through October. We had spent the day at the event, listening to live music and entertaining Kyler. She still had face paint on as she floated around the pool in her water wings, her cheeks striped with the orange and black of a tiger.

I slipped into the pool next to my mother, grateful for the moment of peace. In a few hours, Oktoberfest would close for the day and this pool would be crammed with people, but for now we were the only ones enjoying it. The event had been packed; we could barely move through the crowds. Snowbird can experience some of its busiest days during Oktoberfest, with more customers coming up for the beer on these days than on a powder day.

Many ski resorts are diversifying their business models in anticipation of climate change. One of the ways the ski industry will be affected by a warming planet is shorter winters, and in most locations around the United States the ski season is getting pushed back by days and weeks. Many resorts are starting to struggle to open for the holidays, a crucial time for resorts from a business perspective. Seasons are ending earlier, as well. By expanding their summer business, building zip lines and mountain bike trails or hosting summer festivals, ski resorts are adapting to a changing climate.

My mother smiled at me as I passed her a beer. She's never loved classic beer, but she does enjoy a good grapefruit radler.

"Coming to the Oktoberfest while you're pregnant wasn't great planning," I teased Reyna. She had arrived in Utah with the news she was pregnant again. All bets were on that she'd have a third daughter. There have only been women born into my mother's family for the past three generations.

Kyler was twirling around in her floaties, entertaining herself on the far side of the pool. From somewhere in the surrounding pines a Steller's jay called. I felt a tinge of jealousy that Reyna was pregnant again, that she could make the decision to have children so easily. I imagine that it's an easy decision for most, but whenever I talk with my group of girlfriends about motherhood, the conversations are weighed down with hesitation and fear. We are scared of bringing children into a landscape destined for climate change.

"How did Grandaddy end up investing in Snowbird?" Reyna asked, stretching her arms along the side of the pool.

While my paternal grandfather, Junior Bounous, was helping to design runs and create the ski school at Snowbird, my maternal grandfather, Grant Culley, was donating funds to help build the resort. He was the first monetary investor, and was given the first pick of rooms to own at the Lodge. He chose ones on the topmost floors, a set of four that sits where the hotel bends northwest so that our family could enjoy direct mountain views as well as views down the glacial valley. We were staying in those same rooms that weekend.

"A man named Ted Johnson," I answered. "He was Snowbird's original developer. When Grandaddy and Gummy used to ski at Alta, he was there too and had the idea to develop Snowbird just down the canyon from Alta."

"Wow," Reyna said, "Grandaddy must've trusted Ted a lot to put so much money into an endeavor as risky as a ski resort. You're basically investing in the developer."

"He definitely had a great relationship with Ted," my mother responded, "but the way Grandaddy saw it, he wasn't *really* investing in Ted. He did it for the snow."

Her words snagged in my chest.

"The snow?"

"The snow, and the mountain. He'd been skiing at Alta for a while, and we'd always go into the backcountry. A bunch of that terrain became part of Snowbird when it was developed. So he already knew the mountain, he knew the terrain, and he knew how consistent and light the snow is here. He said it wasn't even a decision, really. He knew that Ted was a reliable guy and a hard worker, but the real reason he agreed to it was because he trusted that the snow would always come. He knew it'd be a good investment from the start."

My grandfather Grant Culley was a businessman who created a life for himself and his family by investing. When he began studying at Stanford University without any money to pay for his tuition, he invested in himself, gambling his way through school. Afterward, he put his time, money, and career into an insurance company that initially failed. When he and his partners tried to launch their business again, it succeeded. My mother was a little girl when Merrill Lynch bought the company, and their stocks tripled overnight. My grandfather then set his sights on purchasing land. Responding to ads in *The New York Times*, sight unseen, he bought a bay in the Inside Passage of Canada and a coconut plantation in Fiji. Our family still owns the cabin in Canada.

"He always wanted to invest in snow," my mother continued. "He was addicted to light powder and steep terrain. When I was a kid we'd always have to drive all the way out from the Bay Area to the Wasatch to ski, rather than just going to Lake Tahoe. And he'd pack up his and Suzanne's ski gear and fly up to the Canadian Rockies to go heli-skiing with a guide named Hans Gmoser. They'd stay in an old logging camp with no central heating or plumbing."

"Gummy would stay in a logging camp without plumbing?" Reyna said incredulously. "Our glamorous Gummy, who had a different ski outfit for every day of the week?"

"I know, I still have a hard time imagining it," my mother responded with a smile. "But within a few years, and with funding from Grandaddy, Hans upgraded and built a lodge in an area known as the Bugaboos."

"But Grandaddy never really saw any return from these investments, right?" Reyna asked. "It's not like our family makes any money from owning rooms here."

"No, but that was never part of it for him." My mother smiled softly, her almond-shaped eyes looking up to the auburn and green hillsides. "He always called it a passion of the heart."

❄❄❄

Before my grandparents were born, others had their own stakes in this same landscape, but for different purposes. In the mid-1800s, gold, silver, copper, lead, and zinc were unearthed in the Wasatch and in the Oquirrhs, the mountain range across the Salt Lake Valley. The discoveries attracted miners to the Utah mountain ranges. The township of Alta was created up Little Cottonwood Canyon to accommodate thousands of miners, along with railroad cars to get the materials and mined minerals up and down the canyon.

The town flourished for decades. It had its hardships, including a fire that destroyed almost all the buildings in the 1870s and a series of avalanches that took over one hundred lives. In their haste to build a town, the residents chopped down much of the old-growth forest, not realizing that the trees had anchored the snow on the steep slopes.

That was not the only foresight the mining population lacked. Mining for minerals is a classic boom-and-bust cycle. The money flows vivaciously when there are accessible minerals in the ground. But mining, like oil, gas, and coal, is not a renewable resource. Once the resource dries up, the money does as well. By the 1920s, there were only a few residents remaining at Alta. One of them, George Watson, elected himself mayor and began buying up old mining claims, hoping that another boom would come around. When it never came, he sold those claims to the US Forest Service. That land would later be integrated into Alta Ski Resort.

Miners relocated their lives to try to strike it rich, but many of them never considered that it was not that great of an investment. They'd flock hopefully to wherever whispers of minerals were reported, settle down, create a town, build infrastructure, and elect officials, never considering that the prosperity wasn't permanent.

A boom-and-bust cycle is happening today, this time with fossil fuels. Many have invested in those industries with the same mindset miners had—that the prosperity from investing in oil and gas is permanent. Fossil fuels haven't hit the bust part of the cycle quite yet, from a monetary and political perspective at least, but they're much more devastating than their mining predecessors. Not just because they will run out, like any non-renewable resource, but also because the carbon dioxide and greenhouse gases they emit into the air will translate to sea-level rise, mass extinction, droughts, destructive storms, and perhaps even the end of society as we know it.

A new kind of migration has influenced Utah in recent decades, one composed of skiers, snowboarders, and outdoor enthusiasts instead of miners. It's a migration that many of my friends are part of, that Colin is a part of. They may be drawn to these locations for recreation rather than extraction, but they have similar mindsets to the miners: this prosperity is permanent.

Warmer temperatures, more precipitation in the form of rain rather than snow, and dust storms will be the downfall of snow in the American West. Climate change will send the ski industry into decline, degrading the snow, which we didn't realize was a nonrenewable resource until now. The demand for snow will grow as the resource becomes scarcer. As resorts in lower elevations close, more enthusiasts will have to travel further from their homes to ski. Higher-elevation resorts, like Snowbird and Alta, will become busier, till they, too, lose their snow. Instead of ghost mining towns, we will have ghost ski towns.

When both sets of my grandparents chose lives based around snow, one from a business perspective, one as a career, they never predicted it would become a nonrenewable resource. My parents, too—my mother a ski instructor, my father a ski racer and coach—never realized that a life built around snow might not be stable in the future. They never had to worry about climate change when choosing a career.

But my sister, Tyndall, an instructor at Alta and a coach at Snowbird, is aware of the risk of investing too much of her life in snow. Not many of her coworkers realize this same risk.

We returned from the pool and I sat on the couch to do some writing. Kyler walked up to me and looked down at the words on the page. Picking up a nearby pen, she underlined a word.

"What does that say?" she asked.

"Snow," I responded, smiling at her.

She underlined a phrase.

"What does that say?" she repeated. I stopped smiling, alarmed at the words she chose.

"Mass extinction."

"And this one?" she asked, underlining a full sentence this time.

Fossil fuel industries are propelling climate change.

She looked up at me expectantly, her round olive eyes soft with curiosity. I swallowed, the words dry in my mouth as I spoke them aloud. She tilted her head to the side, blond curls touching her cheek.

"What does that mean?"

I couldn't answer her. How do you explain climate change to a child?

Waking

I wake up early to get to his house before he awakens. His front door is unlocked, and I push it open softly. His bags from the river are all over the floor, still full. The wooden floorboards creak underfoot as I cross the living room and walk down the hall to his room. He is still asleep, legs tangled in the thin sheets, his hair spiraling out across the pillow in blonde ringlets. I lower myself onto the other side of the bed as gently as I can, but my movements make him stir. Colin reaches over and pulls me closer to him.

"Hi," he murmurs in my ear.

"Hi," I whisper back, smiling.

"Do you want to bike over to my grandparents' house?"

I look up from the dish I'm washing.

"Bike? To your grandparents'?"

"Yeah, it's only ten minutes away. We can stop at the bakery en route and pick up some coffee and pastries."

I haven't met Colin's family yet, besides in passing prior to us dating. The Gaylords have family gatherings in Salt Lake City quite often, but Colin has never invited me to one before.

"I don't have a bike . . ." I start, but he interrupts.

"I've got an extra one you can use."

I busy myself with some dishes to hide my pause. Now that we're both returned from our summer travels, our relationship will have to move in some direction. It had been on the cusp of transforming prior to our separation, but had halted when Colin left for Green River. It seems to me that we can't sustain such a casual relationship; we'll have to either move forward and become more serious, or retreat to being friends again. We've reached a crossroads.

The moment stretches on. If I hesitate any longer it'll become awkward. My stomach churns a little at the thought of meeting his grandparents. And, though I don't care to admit it, biking on streets makes me nervous.

"Do you have a helmet?" I ask.

A few minutes later we're coasting down Colin's driveway. I feel a little silly, wearing a helmet when he doesn't have one on, but I don't quite trust myself to not fall over during this ride. It's a quick trip, even with the stop at the bakery. Within ten minutes we are stepping off the bikes in front of a house with a pile of aspen logs on the front porch and a rack of antlers mounted to the garage.

"Does your grandpa hunt?" I ask, eyeing the rack.

"He lives for it," Colin responds. As we walk up the driveway my ears pick up a familiar song coming from a tree nearby.

Weee-wooo. I turn my head, listening intently. I recognize the birdcall, but can't place the name.

Weee-wooo. Then a series of garbled noises makes my memory click.

"Look, a black-capped chickadee," I say, pointing to the black head and white breast as I see it.

Colin glances over toward the tree, raises an eyebrow, then knocks on the front door, apparently uninterested. My face flushes a little with embarrassment. Maybe I've shown my bird-nerd cards too soon.

Colin's step-grandmother, Claire, opens the door and welcomes us inside. We are greeted enthusiastically by a large black dog—his grandfather's hunting dog, Tru. Bud Gaylord, Colin's grandfather, is sitting at a round dining room table. We sit next to him, placing the steaming cups of coffee down while Claire grabs plates for the cinnamon rolls.

After some questions and conversation about how Colin's summer on the river went, our conversation turns to Snowbird and family.

"I've known your grandparents, Junior and Maxine, from way back," Bud says. "Back when we were building the Cliff Lodge."

"I forgot that you built the original Cliff." The Cliff Lodge is one of the hotel accommodations at Snowbird, similar to the Lodge, where my family has the set of four rooms. "That's where my parents got married, actually."

"Oh, they did?" Bud chuckles, scratching Tru behind an ear. "Ah, it broke my heart when we had to sell the Cliff. We were trying to avoid bankruptcy. We *did* avoid bankruptcy, but letting go of that place was hard. Now, I know who your father is, but who is your mother? Is she from Provo, too?"

"No, my mother's family is from a town near Palo Alto in California," I say. "Her maiden name is Culley. You may have known my maternal grandparents as well—they were some of the original investors in Snowbird."

"Not Grant and Suzanne Culley?" Bud asks, lifting his eyebrows. "We used to have a set of rooms right next to their rooms at the Lodge."

"That's them!"

Bud looks at me over his cinnamon roll, a scrutinizing expression in his watery eyes.

"Does your family still own those rooms?"

I smile. "We still celebrate Christmas there every year."

He nods, his gaze slipping from my face to something behind me, something in the past.

"We used to spend Christmas there, too."

He continues to stare so intently at a memory that I imagine I can smell the scent of cinnamon and pine, hear the music and bells of a holiday long gone.

"Your grandmother had a piano in those rooms, a grand piano—I have no idea how they even managed to get it up to the sixth floor!—and she used to play the most beautiful songs on it."

His hands rise to waist height and he sits up straight, as if there is a piano in front of him. His eyes turn suddenly toward me, with the same lightness Colin's gets when he's excited.

"We could always hear her playing Christmas carols through the wall. Do you still have that piano in the room?"

"We do," I assure him, beaming at his excitement, at the image of my grandmother playing the piano with her family circled around her, singing as she plays.

"Ah, you do. She was quite the woman."

The conversation transitions to December and Bud's upcoming ninetieth birthday. In the next year, Bud will turn ninety,

Colin's father, Randy, will turn sixty, and Colin will turn thirty. Grandfather, father, and son, at thirty-year intervals.

"You guys could add on another little one to the mix," Bud jokes. "Ninety, sixty, thirty, and zero. Four generations of Gaylords, thirty years apart. This would be the only year to do it, better get to it!"

We laugh, but a heaviness lodges in my chest at his words. As silly as it may sound when I say it aloud, the risk of losing a child deters me from even wanting to have one. Any child I have could witness a World War III over access to fresh water, islands and cities disappearing beneath the ocean, and a refugee crisis the likes of which the planet has never seen. She might see the western United States experience a drought that could turn the Salt Lake Valley into a dust bowl. Her children may never see glaciers in their lifetimes. And, with more precipitation in the American West falling as rain rather than snow, my grandchildren may never get the chance to learn how to ski, or to catch a snowflake on their sleeve.

We emit carbon during our lifetimes, leaving our footprint on the earth's climate. It's unavoidable living in the United States, where our buildings and transportation systems and economy are powered by fossil fuels. But the biggest carbon footprint you can leave as an individual is having children. By one estimate from a study conducted by Oregon State University in 2009, "under current conditions in the United States . . . each child adds about 9,441 metric tons of carbon dioxide to the carbon legacy of an average female, which is 5.7 times her lifetime emissions." If that child has more children and grandchildren, that number grows. By choosing to have children, we give choices to others to continue our carbon legacies.

I've had countless conversations with other young women my age who experience this same anxiety. The discussions seem to be geared toward women especially; sometimes it feels like it has become the women's responsibility to save our climate by choosing to forgo starting families. A woman's "carbon legacy" is seen to include more than a man's carbon legacy does. I recognize this, and I know that this conversation is putting emphasis on individuals reducing their personal carbon emissions, when in reality climate change stems from the choices of large oil companies and politicians more than from our individual choices.

And it does not escape my notice that mothers are often the ones on the frontlines of environmental and climate activism. They are the ones who stand up to big oil companies when the health of their children is threatened. Becoming a mother is becoming a fighter, a bird that takes wing when there is a threat too close to her hatchlings to either distract or attack the threat, putting her life in harm's way to protect her chicks. But even so, when it comes to imagining what my daughter and granddaughter's lives will be like by the year 2100, I grow fearful.

Of all the conversations I've had surrounding this topic, I had never thought about how this conversation might go with a significant other who wants children, or with his family who expect grandkids. I become uncomfortable as I sit between Colin and Bud. What if Colin wants to raise a family? If I became firm in deciding not to have children, it could be a deal breaker for us one day.

I stop traffic on the ride back home, trying to follow Colin through an intersection as he turns left. I feel so stupid, the hem of my dress flapping in the wind, threatening to fly up to my belly as I bike in front of the stopped cars. But as we turn down a less traveled road, Colin slows so I can ride next to him. He reaches his hand out to me. Carefully, I lift one of my hands off the handles, seeing if I can maintain balance. In an embarrassingly slow motion, I reach a shaking hand out. He takes my hand in his for a brief moment. Then I wrench mine back again, the bike wobbling dangerously beneath me.

II. Season of Changing Colors

Listening

From the valley floor, Rock Canyon appears rugged. The jagged peaks at the canyon's mouth look like teeth, framing the dramatic *V* and dropping straight into the valley rather than easing their way lower in elevation through foothills. Beyond this imposing opening, however, are the soft slopes of a basin shaped like a crescent moon, tips close to kissing at the mouth of the canyon. On the backside of the teeth are steep but smooth hillsides, carpeted with pines.

It's a glacial basin plush with life, and today it's on fire. Maples, ranging from beige-rouge to brilliant oranges to deep vermillion reds and pinks, grow thick here. While the higher reaches of the Cottonwood Canyons become patchworks of evergreens and golden aspen during the fall, Rock Canyon becomes a rash of ochre and cinnamon from scrub oak and maples.

I stood on the edge of this basin with my grandparents. It was the most delicious of autumn days: warm but with a cooler edge to the air, especially noticeable after an exhaustingly hot and long summer. I had never been here before, though this place is as thick with my family history as it is with maples.

My father grew up on a fruit farm in Provo, Utah, near the mouth of Rock Canyon. The farm was settled by my great-grandfather's family at the beginning of the twentieth century when they immigrated from the mountains of Northern Italy to the mountains of the American West, bringing a knowledge of the land and the family name "Bounous" with them. It was a large plot of productive land that fed the surrounding community for decades, until the city of Provo took the land out from underneath our family through the power of eminent domain. They replaced acres of orchards partially with Timpview High School, one of the largest schools in Utah Valley. Most of the fruit trees, however, were ripped out and replaced with grass. Just grass. My sister, cousins, and I used to wander out of our grandparents' backyard, which still sits on a small corner of the original farm, to go play on this grass. But there's not much adventure to have out there. A small playground, a running track; not the adventure that we would have had if we'd been able to chase each other through the rows and rows of fruit trees. They named this plot of grass "Timp Kiwanis Bounous Park," and gave my grandparents what the city considered "just compensation." But how can you possibly capture the compensation required of robbing an entire community of fresh fruit, stripping multiple generations' work of tending to the land and almost century-old trees, and erasing the fantasies and wonder of children not yet born?

Seventy years ago, when the gnarled trees were still rooted deep in the soil, a rainstorm halted work on that farm. Rather than waiting idly for the farm to dry, Baqui saddled up his horse and rode up Rock Canyon. It was late afternoon when he began his journey. Leaves, vibrant in the golden sun, contrasted against the dark, wet branches of scrub oaks and cottonwoods. The soothing scent of rain on soil stirred within the sultry air currents of the ravine. Somewhere up the canyon he came upon two pants-less women and their horses. My grandma Maxine and a friend had been caught in the rain while riding, and after the rain subsided they lit a fire and took off their pants to dry, never expecting that they weren't the only travelers in the canyon that afternoon. It was the first time Grandma and Baqui met.

At my request, my ninety-three-year-old grandparents loaded themselves into their car so I could see one of the most beautiful places to watch summer turn to fall in the Wasatch. My grandpa agreed to the trip so he could cut another branch of crimson maple to brighten his kitchen. My grandma, her dementia keeping a mild smile on her face even through the bumps of the dirt road, came along for the ride.

We'd already stopped a few times during the winding drive, my grandpa pushing his bent body out of the seat each time so he could point out certain features in the landscape and answer any questions that formed in my mind as I scanned the scenery. If my constant flow of queries about everything in sight annoyed him, he didn't show it.

We summited a crest that would allow us a view into Rock Canyon and the car rumbled to a stop, the tires kicking up dust on a road dried out by the summer. The smell of dirt and earthy rot of fallen leaves wafted through the air as our boots ground the gravel beneath them. The trees seemed to breathe in relief at the upcoming promise of cooler times, the light breezes pulling whispers from their canopies. With the effort of someone experiencing sciatic pain, my grandpa began ambling toward what looked to be the rocky exposure of a little cliff or ridgeline, about fifty feet down a primitive trail.

Pausing, he turned to my grandma and, for the second time that day, advised in his slow voice, "Ah, Maxine, you better stay by the car." And for the second time that day, she responded by moving forward to take his hand and walk beside him.

I followed the two of them as they navigated around trees that seemed strange in this mountain landscape. Instead of dense maples and scrub oak, this little ridgeline was scattered with flora that looked drier than the rest of the vegetation in the basin. I recognized the soft, peeling bark of junipers, more common in the Utah deserts than in the Wasatch, but couldn't name the other tree.

"Baqui, what kind of tree is this?"

"Greasewood. Well, that's what we call them at least. Their bark burns black."

Dry, golden grass brushed our ankles and calves as we stepped over the crumbling dirt trail. Our destination was an outcrop, a place where bedrock is exposed. Outcrops aren't covered by soil and vegetation, making them easy features for geologists to examine without having to excavate. When we look at outcrops, we see the bones of a landscape.

This outcrop is where my father's family had "weenie roasts" every summer and fall. Though he admitted the official name might be something else, my grandpa called it Rattlesnake Ridge.

The name jogged a memory. I already knew this place from a story my mother told me. My dad invited my mother to this place on one of their first dates to join his family for a weenie roast. According to my mom, Baqui, already over fifty years old at the time, started at his house and ran up Rock Canyon to meet them, rather than riding in the car.

I brought up this story and my grandpa chuckled.

"I just dropped your Uncle Barry off here the other day so he could do that run in reverse."

During the weenie roast, perched on that rocky outcrop, my mother witnessed the underlying bones of my father's family, what lay beneath, what made their family what it was: a deep and desperate love of the mountains, of the soil and the maples and the dry, oxygen-deprived air. She also glimpsed what her future could be as a part of that family, what a relationship with my father might become.

We emerged on the outcrop and my grandpa made a sweeping motion from the tip of the peak to the east, known controversially as Squaw Peak, tracing its shoulder down as it arced toward us and then south, dropping into the basin. He explained this unseen boundary as the northern edge of what used to be the homestead of a man named Louis Richard. I'd never heard his name before, and was startled when my grandpa said that Louis was his uncle, his mother's brother, a shepherd who used to bring his flock into Rock Canyon during the summers in the early 1900s.

My grandpa pointed to a part of the hillside across from us that looked a little different than the rest of the basin, like an old scar on a patch of skin. During one particularly violent summer rainstorm, Louis heard a loud rumbling coming from somewhere above his cabin. Trusting his instinct, he jumped on his horse and rode as fast as he could out of the canyon. A landslide followed him all the way out. My grandpa remembers visiting the cabin back before the mudslide destroyed it, when he was only ten years old. Later, Louis had to sell his homestead to the Forest Service, adding another few hundred acres of land to the Wasatch-Cache, adding family lore to the quilted landscape of the mountains.

Listening to my grandpa's stories, I felt as though a big sewing needle was sewing my body into the fabric of that basin, that a part of me I hadn't realized existed was being exposed to the mountain air like the bedrock we stood on. Here was another place that made up the landscape of my grandfather's family. An exposed outcrop where I was able to glimpse the past, the layers that my family and I are composed of.

Author Kathleen Dean Moore says that "we're born into relationships, not just with human beings, but with the land—the beautiful, complicated web of sustaining connections." My children will be born into a relationship with this land, just as I was. I imagined bringing my children and grandchildren to this very ridge, pointing out the features, naming the trees, telling them stories of their great-great-grandfather and his uncle who used to have a homestead here, recounting how the first hints of love began between their grandparents during a weenie roast.

And then, the tinge of remorse that comes when thinking about starting a family.

If I have children.

Standing there on Rattlesnake Ridge, I became suddenly aware of my blood moving through my body. It was the same color as the maples and oaks around me. A silent word echoed in my mind, bouncing off the walls of the basin and ringing in my ears as the sound of grasshoppers and finches, filling my nostrils with the scent of dust and leaves.

Home, they said to me. *This is your home.*

Mobbing

"Oh no." I take a shaky breath, gripping the wheel. "Oh no oh no oh no."

"Easy now," Colin says from the passenger seat. "Just go slow."

The steep decline had turned to mud. The typical pink sand of southern Utah is a sultry shade of burnt sienna. Heavy rain and hail in the past hour eroded the right side of the road and created gaping cracks, some of them quite large, in what should be a dry dirt road down to the campsite. If the car starts sliding there is a good chance we'll drift off the side and get stuck. There is little room for error.

We were finishing a hike in Zion National Park when the sunny afternoon turned into a raging hail and lightning storm, leading to flash flood warnings and a mass exodus from the park. As we boarded the bus to leave Zion, three ambulances with their lights on passed us going up the canyon. I fidgeted in my seat nervously, thinking of the riverside campsite we had planned on staying at that night.

I tap on the brakes lightly, and we creep slowly down the hill until it flattens out. We step out of the car, our hiking boots sinking two inches into the mud. I feel as though I'm walking through a marsh, not a desert. Colin starts looking for the highest place in the campsite, where we might be able to set up a tent. I watch the river. We had been looking forward to our first night camping together, but rain in the desert makes me nervous. The slickrock of southern Utah doesn't absorb water like soil does. Flash floods are frequent here. I pick out a few exposed rocks in the flow of the water and watch them. Within minutes the river rises and the rocks disappear.

I hold the keys out to Colin.

"I don't want to stay here," I say, then, after a moment's hesitation, add, "I'm a little scared."

"Fine by me," he says, taking the keys, recognizing that he'll be responsible for getting us safely out.

"The mud's too thick to camp in anyway."

The sun is setting as we climb back out of the campsite. Dark, dense clouds line every horizon, hovering over scarlet cliffs, but there's a break in the clouds directly above us. The last of the sun's rays graze the sky, golden and bright.

"Look!" Colin says suddenly, hitting the brakes and pointing to a telephone pole on the side of the road. There is a giant owl on the pole, iconic, tufted horns on its head. Another owl is in flight, enormous wings outstretched, the white plumage golden in the setting sun, wingtips dark. It flies at the one on the pole, extending sharply silhouetted talons. The owl on the pole rises in flight at the threat, great wings lifting the dense body into the air, talons extending in kind. They both disappear over the roof of the car.

"Those were great horned owls!" Colin says, stopping the car completely. He opens the door and steps outside, looking to the sky. I follow, but the owls have already flown into the shadows of nearby trees.

"They were mobbing! I can't believe it, my dad's going to die when I tell him," Colin continues, laughing. "On the fall equinox, too. Wow." He stares at the trees they had disappeared into.

"You're into birds?" I ask.

"Yeah, sort of. I mean, my dad is. We once flew in from Washington to do a HawkWatch backpacking trip with my grandma, west of the Great Salt Lake." He shakes his head as he gets back in the car.

"I've only ever seen a few owls in the wild before. To see two great horns, mobbing at sunset? So crazy. Their wingspans must've been, what, four, five feet across? Did you see their talons?"

"Yeah," I say, smiling as the car rumbles back into motion. "I'm kind of surprised, you never seemed that interested when I pointed out birds before."

"That's because you only like the little birds," he teases.

Adapting

One brisk October morning, my parents and I summited Mount Timpanogos. The hike is about fourteen miles and almost ten thousand feet of elevation gain and loss. My father and I were going to make a loop out of it, beginning at the Timpooneke Trailhead and finishing at the Aspen Grove Trailhead. My mother would split off at the summit and return to Timpooneke to pick up the car. I was not disappointed; the peak was gorgeous, with 360-degree views of the glacial basins and cirques of Mount Timpanogos, Utah Valley, Utah Lake, and the Oquirrhs.

After summiting, my father and I descended along the south shoulder of the mountain. We quickly lost the trail and got cliffed out. Our options were to turn around to try to find the trail we missed, which could be a mile of backtracking, or downclimb the fifteen-foot cliff, which would take us to the proper trail. My father was certain we could figure a way down.

He climbed first, stopping about halfway when he reached a small ledge. If either of us lost our footing, it'd be a nasty, even deadly, fall. But I had followed my dad around rocks before and was confident I could follow him again. I was his eldest daughter after all, his adventurous spawn who'd blindly follow him off a cliff.

Before I was even born I had a legacy of mountain climbing. My mother had three miscarriages before me. Each time she became pregnant she'd follow the advice of her doctor, taking it easy, not participating in strenuous activities. And each time she'd lose the pregnancy. She and my father had already planned a climbing trip to the Bugaboo Mountain Range in British Columbia, Canada, when they got pregnant again. She started spotting while they were there, but instead of taking it easy like she had for her first three failed pregnancies, she decided she wasn't going to let this one hold her back.

Their plan was to summit a peak called Bugaboo Spire. Halfway through the day their guide bailed on them, claiming that the weather looked too bad to summit. My parents stayed on the mountain, discussing if they wanted to retreat or not, and while they were deciding, the weather cleared up. My father thought that he could get them up the peak, so they decided to go for it. He was elated when they made it to the top, but my mother had learned from years of following my father that the hardest part about climbing is the descent. She was so nervous she couldn't eat the lunch she had packed. She was experiencing a profound sense of responsibility, starting to comprehend that at three months pregnant she was "in it for two."

During the descent they had to rappel down the back of the spire, three thousand feet above a glacier, then walk across an exposed ridgeline with no protection. My father tied the two of them together with a long rope, so that if she stumbled and fell he could catch her.

"What if *you* fall?" my mom asked him skeptically, eyeing the rope that connected them.

"Jump the other way so our weights will balance out and catch each other."

"What if I accidentally jump the wrong direction?"

"Well, honey, then I'll see you on the way down."

My parents gave me the nickname "white-knuckled womb-clinger" before I even made it to the second trimester. I was the first pregnancy to make it, the one that held on.

The top section of the cliff was easy, slightly slanted so I could clearly see where to put my feet. The final part of the climb was more difficult. The wall became a slight overhang so we couldn't see what was beneath. My dad was a confident enough climber that he could hang on with his hands while his feet sought a hold, but I was not. He lowered himself first, stopping a few feet above the ground, then instructed me on where my feet would move next. I was shaky and sweaty even before I started climbing. I lowered myself one hold, then another, but I couldn't reach the next hold. My foot dangled in the air. My fingers were holding onto a slanted rock, and they were slippery with sweat.

"Just another few inches, Ayj, you're almost there," my dad said.

"I—can't!" If I hadn't been so terrified, I would have been mortified by the way I shrieked at him. My fingers were slipping. The foot that was still on a hold was shaking so badly I thought it might shake itself off the wall. I was going to fall. And I was going to take my dad down with me, and we'd crash onto the scree field beneath us and tumble down the side of Mount Timpanogos.

Tears began to blur my vision. I tried to pull myself back up, but I had lowered myself too far to even do that. My knee was practically in my armpit, and I didn't have the strength or the wit to stand up from that position.

"No, no, don't go up now! Ayja you have to trust me, your foot is almost there. Here, I'm going to guide it. Take a breath and exhale."

My dad let go of one of his holds, reaching up to grab my shoe. He began pulling it down, and I knew I was going to fall. Involuntary whimpers slipped from my mouth. There was no way I could lower myself another inch. But even as I thought it, my body somehow found that inch. My foot touched a hold.

"There, you've got this."

And then we were off it. As soon as the adrenaline faded, embarrassment set in. It was so potent I thought I might start crying again. Why did I have to lose my head like that? My father had gotten me into situations I felt uncomfortable with before, and he'd always gotten me out of them. He may have tortured my mother during the earlier days of their courtship with some of his adventures, but he had mellowed out since then. I thought that I could finally fulfill the role of the fearless offspring of Steve Bounous, but I was wrong.

We continued onto the saddle, stopping before we cut down onto the snowfield and rock glacier to appreciate the view, which looked into two enormous cirques. I sat down on a rock, looking up at the steep hillside and boulder field behind my father. It took a moment before I registered what I was seeing.

There was a mountain goat twenty feet from my dad, the color of freshly fallen snow. It seemed massive standing so close to him, shaggy and burly, with horns curving backward from its thick skull and a beard dangling from its chin. Its dark eyes were focused on my dad, who only noticed it when I said something. It blended in with the side of the mountain, but once my vision registered it I saw two more just behind it, and another three to the right of them.

The hillside came to life. We counted fifteen, twenty, twenty-seven, thirty-five, forty-three—the more we looked, the more we saw. We couldn't believe we hadn't noticed them earlier. Most of them weren't even standing still, but were moving around the jagged outcrops. I guessed that about half of them were kids. They were especially active as they followed the adult goats around, testing their young hooves on the rocky paths. I had never seen mountain goats at this distance before. They were so sure-footed, prancing up rock faces that looked smooth. I became jealous as I watched how confident the kids looked as they chased each other across the ledges. No shaky hooves there. We watched them for half an hour before continuing down the snowfield.

Mountain goats are hardy creatures. Kids will start walking and climbing within a few hours of birth, following their mothers, or nannies, closely for a year before becoming more independent. Nannies and billies are protective of their space and food sources, lowering their horns and fighting other goats. Because they mostly stay above the tree line, they can avoid potential predators like wolves and bears, but cougars and even golden eagles still remain a threat in high-altitude cliffs. Golden eagles will target kids, diving and grabbing with talons to pull them off the wall and let them fall to their deaths.

Goats are well-adapted for life in the Wasatch. Their hooves are sharp and precise, allowing them to access and stay on ledges hundreds and thousands of feet tall. My dad was once leading a climb up Little Cottonwood Canyon called Pentapitch. He was on the last pitch, about five hundred feet off the ground, when a few rocks hit his helmet. A mountain goat was perched on a ledge a few feet above him, staring down at him. For a few moments, mountain goat and mountain climber stared at each other. Then the goat lifted up a hoof and knocked a few more rocks off the ledge. My father couldn't travel sideways in that particular spot, and he was far enough above his last piece of trad gear anchoring him to the rock that he didn't want to downclimb. So he waited on the wall, keeping his vision down in case the goat knocked more rocks on him. The goat moved when it realized my dad wasn't going anywhere, climbing further up the side of the cliff.

Like many alpine species, mountain goats might be some of the first to feel the effects of climate change in high elevations. They move with the snow, retreating into the highest reaches of mountain ranges during the summer and venturing into lower elevations during the winter. They rely on high alpine vegetation for survival: mosses or lichen that grow in the highest elevations or grasses found in lower alpine meadows. As climate change impacts how vegetation grows in the Wasatch, warming alpine ecosystems, species that historically grow in lower, warmer elevations are moving higher in elevation, outcompeting current vegetation. Trees and shrubs are encroaching on alpine meadows. As this occurs mountain goats might be forced into higher elevations where there is less food available, and will have to compete with each other more for resources.

We slid down the snowfield, careful not to pick up too much speed while glissading. My father searched until he found a place where we could cross from snowfield to boulders without having to jump across a gap, like he made my mother do so many years ago. Even so, hiking down the boulder field was precarious. I took care as I stepped across rocks, all too aware that a misstep here could mean a broken ankle or worse.

Occasionally we'd hear the *Eeee!* of a pika during our descent, or see a pile of grass among the rocks, where a pika had been gathering food for winter. Pikas are small rodents that live in rock debris like scree and talus. They look similar to mice, with golden fur, large ears, and whiskers wider than their bodies. Pikas live at high elevations, preferring altitudes around nine thousand feet. Unlike many alpine rodents, pikas don't hibernate but instead dig tunnels through the snow and remain active during the winter. They have thick fur and are very sensitive to changes in climate. They can die from heat exhaustion at temperatures as low as seventy-eight degrees Fahrenheit, but they can also freeze to death if the temperature drops to fourteen degrees Fahrenheit. Temperatures in the peaks of the Wasatch can drop well below fourteen degrees during the winter, but pikas are able to survive even the most frigid temperatures because of the insulating qualities of snow.

Heat, which is essentially molecules and atoms vibrating quickly, has a hard time moving through snow. In order for air to warm up, the molecules in the air have to be able to move. Because miniscule air pockets in snow are trapped by the structure of snowflakes, they don't have much wiggle room to heat up, and it can take a long time for heat to transfer through snow, especially freshly fallen snow with lots of air pockets. A layer of snow helps prevent heat from escaping spaces, which is why hibernating animals and some human populations dig into the snow to create homes. Snow at the bottom of a snowpack keeps the ground warm as well, which protects seeds, plants, and animals from the frosts of the air above.

For pikas to survive winters, they need a thick snowpack that they can tunnel through, which shelters them from cold temperatures. During the summer, they need cooler air currents and rocks to protect from heat. Climate change is threatening pikas from all sides. Summers in the mountains are becoming hotter while the snowpack is diminishing in the winter. Since they can't move into lower elevations due to heat, when pikas are under stress they'll move higher in elevation. Eventually they can't go higher. Pikas are already locally extinct in many parts of the American West, and are considered an indicator species when it comes to climate change and mountains.

Another species experiencing the effects of climate change is the yellow-bellied marmot. Marmots suffer much like pikas do with climatic changes. Their thick coat means they overheat easily when temperatures get too warm, but can freeze to death during the winter if they don't have a thick enough snowpack. Unlike pikas, marmots hibernate during the winter, and the snow insulates their bodies while they sleep.

A motto often used in regard to species and environmental changes is: "adapt, migrate, or die." In parts of Europe, marmots and a certain species of mountain goat called chamois are adapting to climate change by shrinking in size. Smaller bodies are a way of dealing with heat; bigger bodies mean more surface area and warmth. But pikas, which are already so small, aren't adapting by shrinking. And there is no evidence that goats and marmots in North America are adapting like their European counterparts. Migration isn't a reliable option for these species either; they can only migrate so far up the side of a mountain before they reach its peak. So as climate change affects the mountains of the American West, the only other option might be to die.

Flooding

My family once traveled to a small mountain village on the French-Italian border called San Germano, where my great-grandfather and Junior's father, Levi Bounous, was born. For one week we stayed with distant relatives who share our last name. They welcomed us to their home, a beautiful yellow cottage amid a lush green valley. On our first day, they took us on a walk through San Germano. We ambled along the cobbled roads as a pack, eating blueberries their son had picked from their garden that morning. We arrived at the crest of a steep hill, where a fountain had been cut into earth, and we each drank from the spigot. They pointed to a humble two-story house across the street, with views into the valley.

"This is where Levi was born."

The day we arrived was the one-year anniversary of their grandfather, Carlo, passing away. He was the equivalent to Baqui in my family. To celebrate his life that evening, Carlo's children, Loradonna and Mauro, opened up a few of the last bottles of his homemade wine and juniper schnapps, made with ingredients collected from their garden and the surrounding woods. I took small sips of the sweet and syrupy mixtures, feeling as though I was drinking my relatives' love for their mountains into my body. The drinks tasted like summer in the mountains, like soil and pine and citrus. We toasted to Carlo's life, the gesture a cocktail of love and family and landscape.

In the days that followed, I felt an intense connection to these people I had never met and for these mountains I had never seen before. We drove up the winding roads, admiring the crumbling fortresses on distant ridgelines. Peaks loomed further up the canyon, beige and pewter in the higher reaches where vegetation couldn't grow, a bright shamrock green where trees grew thick. I turned to Tyndall as we hiked along a rocky ridgeline. A hawk circled in an uplift, eye-level to us.

"Do these mountains remind you of the Wasatch?"

On our last day there, the Italian Bounouses drove us further up the canyon to their local ski resort, Prali. Dressed in hiking clothes, we rode the lifts to the top of the peak before hiking down into a back basin, one that resembled another basin at Snowbird, where copper water flashes in fields of emerald moss. Named the "Valley of Thirteen Lakes" for the small lakes scattered around its palm, this basin had an old wooden cross at the top of one of the peaks.

Mauro and his son, James Junior Bounous (named in part after my grandfather) pointed out rocks with soil that had been dug out from underneath, signs of marmot life. Wilma, Mauro's wife, touched my shoulder whenever she saw a wildflower that she wanted to show me, naming them and acknowledging which provide remedies for colds and headaches. The flowers there were different than the ones in the Wasatch, but they created the same quilt of colors across the basin—cream and butter yellow and lilac, with the occasional burst of crimson and cranberry. They talked about Carlo often, about how he used to know the paths of these mountains like they were imprinted on his heart.

But there were stark differences to this place. The most prominent were the old army barracks and the two long-range cannons that reminded us that this region wasn't always as peaceful as it was on this day. The ridge that circles the basin is the border of France and Italy, and one of the locations where skirmishes occurred between the two countries during the Battle of the Western Alps in World War II. Originally built in the late 1800s by Giuseppe Perrucchetti, an Italian general who sought to strengthen Italian forces in the Alps by creating what would later be called the "Guardian of the Frontier," the barracks are still standing. The barracks' original rockwork are structural mosaics in that mountain fortress. Mauro pointed to large, flat rocks that scattered the ground around one of the ruins. Chipped into them were names and dates of people who spent time living in the barracks, many of whom were part of the 149th Guardian of the Frontier Artillery Battery stationed there during WWII. Carved into one of these rocks, catching the shadows of the afternoon sun in the imprints, was the name "Bounous."

There may not have been an active war occurring on European soil while we were there, but there was another type of extreme violence. The Syrian refugee crisis was forcing hundreds of thousands of refugees across the Mediterranean, all attempting to escape a country shredded by climate change in the disguise of civil war. Climate change might not yet be creating the death tolls that WWII was responsible for, but it's creating a type of less obvious violence, or what Rob Nixon, an author and professor at Princeton University, calls "slow violence."

Climate change is already affecting Europe. The famous alpine wildflower associated with Austrian independence, the edelweiss, is being forced into higher elevations by warming temperatures. Glaciers are clinging onto peaks and summer ski camps closed due to rotten snow from the higher temperatures. New strains of grapes are being tested in the Alps in preparation for a climate that will decimate the wine country's productivity. Climate change might not be leading to the staggering deaths that D-Day was responsible for in a single day, but it manifests itself in human populations in other ways, like the drought that jump-started Syria's civil war and the searing heat wave that killed over seventy thousand Europeans in the summer of 2003.

While watching a weather clip declaring Kuwait City as the hottest city on earth at over 120 degrees Fahrenheit, my father, without realizing the weight of his words, remarked, "I would be angry all the time if I lived there."

I walked around the remaining cannons in the Valley of Thirteen Lakes, peering down their long dark barrels to try to get a glimpse of an eighty-year-old violence, wondering how long it would take the population of the world to realize that we are already in another world war. It's a different kind of war. The weapons are oil rigs and tar sand mines. The generals and military commanders are the extractive industries. The foot soldiers are infrastructure and automobiles and beef farms. The enemy is ourselves.

Shifting

"I'm assuming you'd like to sit inside?" the server asked as he picked up two menus.

"Actually, could we sit at one of those by the railing?" my mother asked, pointing to a table that was right at the edge of the deck. It had brilliant views of the misty mountain landscape this restaurant resided in, but no awning or umbrella over it. "We want to be able to see the trees."

"Uh . . ." The server wasn't quick to hide his surprise. "Uh—yeah, if that's what you want, that's fine."

Dark clouds marched across the sun, making it brilliantly bright one moment, dark and foreboding the next. It was mostly sunny in the valley, but as we drove up Big Cottonwood Canyon on the way to the restaurant we passed through microbursts of rainstorms. It had stopped raining as we parked the car, but it seemed like a temporary lull. I would have chosen a more sheltered seat, but my mother was optimistic.

The two servers tried to wipe off the table and chairs as best they could. When the chairs were still wet after they finished, they grabbed some sheepskin from inside so our pants wouldn't get wet. My mother sat down—ordering two glasses of chardonnay for us as she did so—and let out a satisfied sigh, smiling up at the mountainside in front of us.

Dense pine forests smothered the mountainside before us, darkened by the rain. Tips of trees were lit up softly from behind by the changing light of the sun, creating feathery textures of lighter gray within their dark ocean. Aspen, sometimes individuals, sometimes in small groves, interspersed the pines. Unlike their evergreen neighbors, their golden coin-shaped leaves became fluorescent when the sun shone through their paper-thin veins. The moisture was so consistent as we drove up the canyon that the air shimmered, the southern rays of the sun touching each descending raindrop so the mountain appeared to us through a lace veil, a waterfall of misted light.

Our server brought out the wine along with a trout appetizer, and my mother tipped her glass toward me.

"Cheers."

"Cheers. To ten years cancer free."

My mother smiled as our glasses clinked. Though the smile reached her eyes, I could still see the pain lodged deeply within them. Pain reminiscent of months of chemotherapy and radiation.

The wind wove an invisible path through the leaves of the nearest aspen. I always thought aspen leaves looked like sequins when I was a little girl, the way one side would be a slightly lighter shade than the other side. I used to draw patterns across any sequined fabric I came across, weaving my fingers through the sequins to make them flip one way and then the other way, their shades of color varying depending on which side was facing up.

Setting her glass down on the table, my mother said, "I hate to say it, Ayj, but I think you'll need to start getting breast exams next year."

Our eyes held each other's for another moment. Busying myself with a piece of trout, I tried to keep my voice casual. "Well, if I am at higher risk at least I've had a good run."

The joke fell flat. In the silence that followed, the soft whispers of quaking aspen sounded like a slow, stirring sigh.

When my mother was diagnosed with breast cancer, I was eighteen years old, on the cusp of leaving for California to go to college. In the weeks leading up to my departure and my mother's mastectomy, only a few days apart from each other, I developed a yearning to re-familiarize myself with the mountains, which I had lost interest in during my teenage years. Perhaps it was the realization that I was moving to a relatively flat landscape, and accessing mountains would no longer be so easy. Maybe I wanted time to reflect on my departure, my necessary separation from my family in order to try and become my own person. Or, more likely, it was to escape a household constantly coping with an autoimmune disease and now cancer.

When my sister was diagnosed with lupus, my mother blamed herself. The doctors have never been certain if it was genetics that caused Tyndall's illness, or outside, environmental factors that affected her body when she was young. Some doctors believe that going on the birth control pill at a young age made her already estrogen-dominant body tip over a threshold during puberty, igniting the disease. Though we never knew what exactly brought it on, my mother blamed her own genetics, also estrogen-dominant, for bringing about her child's illness.

What's most likely is that an array of factors, genetic and environmental, contributed to both my sister's disease and my mother's cancer. Bodies stacked with estrogen, inherited through the veins of genetics, are already weakened to any number of poisons throughout their lifetime. Exposure to toxins, whether through medication or apparently harmless things like plastic and perfume, can tip already stressed immune systems. Those same estrogen-prone roots that my mother passed to my sister are in me.

I would drive up into the mountains, thinking about my body. I was trying to become more independent from my family, yet my skin, my organs, my bones, my hormones didn't seem to be my own. They were never my own, they were inherited from others, from my surrounding environments. Any cancers, any diseases that plagued members of my family could plague me as well. I'd hike through a grove of aspen, each tree appearing individual to the eye, but I knew they were all tied together through their roots. When environmental stresses, like drought or disease, weakened one of them, it would weaken all of them.

Though aspen appear as though they're individual trees, aspen groves are clones of one original tree. Aspen can reproduce through seeds, like most trees, but they are more successful when they reproduce by root sprouts. "Suckers," which sprout off of lateral roots, break the surface of the ground and grow into a new tree. An individual aspen is relatively short-lived, commonly living to 150 years. An aspen clone can live thousands of years. The oldest and largest aspen, and organism, on earth, is in Fish Lake, Utah. It weighs fourteen million pounds, is over one hundred acres large, and is estimated to be eighty thousand years old.

Aspen thrive in disturbed areas. When fire or disease clears out sections of an established forest, aspen fill in the negative spaces between pines. Their colors are a vibrant contrast to their evergreen neighbors, green in the summer, gold and apricot in the fall. Some clones will change sooner than their neighboring clone, or turn a different shade of color, creating mosaics across landscapes. A grove of aspen still completely green can stand next to a grove that has already lost most of their yellow leaves; a grove of amber may hug a grove of persimmon. The openness of their communities creates some of the most biodiverse forests in the West, supporting more wildflowers and wildlife than coniferous forests.

They are sensitive trees. They are sensitive to shade. Once more shade-tolerant species move back into an area, they out-compete aspen until fire reopens a forest to light. They are sensitive to touch. Their soft, white barks became canvases for Basque shepherds during the 1800s, who carved designs in them that we now call *arborglyphs*. They are sensitive to drought. When there isn't enough water in the soil and when the air around them is hot and dry, aspen can experience stress through their veins, akin to a heart attack in humans.

Like all trees, aspen pull water from the soil into their roots and vascular systems. Tubes called *xylem* transport water from their roots into their stems and leaves. At the same time, the tree pulls carbon dioxide, CO_2, from the atmosphere into its leaves. When sunlight is absorbed into chloroplasts—cell structures embedded in leaves—the energy from that light is used to deconstruct the H_2O and CO_2 molecules. The hydrogen, oxygen, and carbon get rearranged into different molecules: O_2 (oxygen), H_2O, and $C_6H_{12}O_6$ (a carbohydrate molecule more commonly called *sugar*). This sugar is then converted to starch, used to create cells, or transported through the organism via tubes called *phloem*, traveling from the leaves down to the roots to assist in plant growth.

The transfer of water through xylem and sugar through phloem is made possible by a vacuum. Tiny openings in the leaves, called *stomata*, are what take in CO_2. But when they open, O_2 and H_2O are released through the stomata. The loss of water creates a sort of suction, a pull which continues to draw water up from the ground, like a sponge continuously losing water and sucking more in. During a drought, when there's little moisture in the soil and little moisture in the air, water will keep being released through the plant's stomata. Higher-than-normal temperatures and dry air stimulate higher rates of water evaporation from tree leaves. This creates more suction, and the need for water in the plant becomes greater. But a lack of moisture in the soil creates a tension within the xylem, stressing the tree's vascular system. That tension can lead to little bubbles of air entering the xylem, like a straw sucking up air when there's no more liquid, which gradually blocks the movement of water through the tree altogether, causing the tree to die.

Trees can close their stomata to prevent the loss of water, but in doing so they are no longer able to take in CO_2, and can't make sugar for storage, building, and energy purposes. If this occurs within an aspen grove, the larger, more established trees might pull sugar from the new suckers, or root sprouts, preventing new growth in a grove. When aspen are under stress, they become susceptible to disease and insects, which continue to damage already weakened groves and can lead to tree mortality.

After a meager winter, when less snow falls on the landscape and quick warming leads what snow there is to melt quickly, dry soil contributes to this stress. Couple that with hot, dry summers fueled by climate change, and my generation might be one of the last to watch how aspen leaves flicker and fly and move like sequins in the wind, or walk through the patches of sunlight that fall between their gnarled white bark within a grove.

It was a Stellar's jay that finally cut through our silence. Following it's screech I glanced around, catching a flash of bright blue from one of the nearby pines.

"CO_2 levels were at 406 ppm today," I finally said to release the tension. I regretted my choice of conversation even as the words formed in my mouth. The number was significant for this time of year, and depressing; usually early fall is when carbon dioxide levels in the atmosphere are at their lowest, since plants and trees in the Northern Hemisphere, which has more flora than the Southern Hemisphere, have spent all summer taking

CO_2 in through their stomata and converting it into cell structures. Once the leaves start falling and decomposing, all of that CO_2 is released back into the atmosphere. To have the lowest CO_2 reading of the year clock in around 406 ppm was alarming.

My mother didn't say anything immediately, taking a long sip of wine. A stronger breeze whipped through the aspen leaves, unhinging a few that took flight. They flickered gold in the afternoon sun, coins tossed into the air.

"You know that Fred stopped by our house the other night?"

I nodded. He was an avid skier and outdoorsman, and the last time we had seen him he had told us how he was voting in the upcoming political election.

"Well, when he came over he was going on about how doomed our climate is. He doesn't seem to think that humankind will last more than a few hundred years."

I snorted derisively. "Did you ask him why he voted the way he did, then?"

My mother shook her head, not as an answer but in disbelief.

"He said that the Paris Agreement wasn't nearly enough, and nothing that we do now will stop climate change."

"Easy enough for him to say. He'll be dead before things really start getting out of hand."

"But he has grandchildren," my mother said. "I asked him how he could talk that way when he has family members who will still be on this planet by 2100. He didn't really give me an answer. He just made it seem like there's nothing to be done."

"How did you respond?"

"I called him out for his pessimism," she said. "I told him that's no excuse to quit fighting. It's like being diagnosed with a terminal illness. Even if the doctors tell you there's no chance of surviving, there's no way someone would just give up all hope and not do anything. I told him that we don't have a choice—we have to keep fighting. We have to choose hope over despair."

Burning

Almost all the leaves in the Wasatch had fallen from tree to earth. My family left the Salt Lake Valley to travel to the township of Portola Valley, in California, where my mother grew up. My grandmother had taken her last breath that prior summer. We decided to hold her celebration of life the following fall in the home she and her husband had built together and raised their family in, and where they had both passed away. Though other members of our family were with her to sing her into her final sleep, my immediate family was as far away as we could be.

My parents, sister, and I were traveling together in Europe when it happened. After spending a week with our distant relatives in Italy, we were staying in Zermatt, Switzerland, for a ski camp my father was running. It was the first camp he had held in Zermatt in ten years, though my sister and I used to spend a month there every summer as children. My grandmother often joined us. While my parents would coach or ski or embark on a twenty-mile trek, she would teach us how to watercolor paint in a little chalet, which we returned to that summer.

Though she'd been declining for the past few years, my grandmother passed suddenly. We only had a few days' warning to hear that she had taken a sudden plunge for the worse. A few hours before she passed away, I walked into a room and smelled watercolor paint, though none of us had brought watercolors. That night, the Matterhorn was draped in robes of brilliant pink—my grandmother's favorite color. She took her last breath in California, but she shared this landscape with us one last time in those moments.

My mother turned the rental car into the long driveway. The custom-made silver mailbox still stood at the edge of the road as it always had, with the name *Grant B. Culley Jr.* perched on the top in melded silver letters. The driveway was lined with small sequoias and redwoods that my grandparents had planted sixty years ago, and their shadows created puzzle patterns across the hood and windshield as we passed underneath them. The car spooked two rabbits foraging near the pavement, and they dashed across the road in front of us, disappearing under nearby bushes.

Palm trees and pines towered over the auburn shingles of the low-lying house. Near the double front doors, enormous bird-of-paradise shrubs were on the brink of blooming, slivers of bright orange peeking out of their green spears. Beyond them, the giant agaves, up to eight feet across and equally as tall, clung to the hillside between the driveway and the breezeway, their sage-colored, swollen leaves twisting out and up to the palms above. Begonias were just starting to lose their color, hints of brown touching the edges of each magenta petal.

Our footsteps echoed in the cool, narrow space of the breezeway between the driveway and side entrance. We glimpsed the backyard, the patio and terraced landscaping, leading down to a turquoise pool and the valley beyond. My mother unlocked the heavy wooden door and pushed against it forcefully, swinging it inward.

We had been expecting this weekend for the past few months, looking forward to it even, but now that the moment of goodbye was here it had arrived too soon. We came to this place to say goodbye to my grandmother, but we were also saying goodbye to her home—the adobe walls my grandfather built by hand, the orchids and poppies and jade, the cracking and sloping tennis court, the orange trees whose fruit my grandmother squeezed fresh juice from every morning, the Cold War-era bomb shelter that my grandfather turned into a wine cellar beneath the house.

We sold it to a neighbor, my mother's best childhood friend, Lynnie, and her husband, Rich. Somehow, this made the sale both easier and harder. They graciously gave us six months to clean out the house before they took over, so we could keep the atmosphere of this home that she loved so much for her celebration. But this house would be demolished within the year.

The day before the rest of our family arrived, my mother, sister, and I took a break from the house. Throwing on walking shoes, we passed through the shadows of sequoias and redwoods. A thick, high canopy of oaks closed in above our heads, cathedral-like with their tall, strong trunks, sheltering us from the midday sun. The noises from the street became muted, and then disappeared altogether, replaced by the distant music of birds above us and the occasional rustle of a bush.

We followed the path of the dry ravine, hillsides rising steeply on either side. Behind Utah in the seasonal cycle, leaves were just starting to scatter themselves across the wide trail. They still had enough moisture in them that they didn't crunch underneath our shoes, but created a thin padding of layers over the packed dirt. There was a coolness to the currents of air that hinted at the shifting seasons. A few days later I would bring Colin on this trail and we would encounter a committee of turkey vultures perched silently on fallen logs across the ravine, watching us through the veils of their dark eyes.

"You know, I always knew I'd have to say goodbye to Mother one day," my mother said, her voice small but strong next to the trees. "That's not the hard part about this weekend."

My sister and I shared a glance as we carefully passed a patch of poison oaks, the edges of their leaves crimson and ominous. We had had this conversation with her many times, but we didn't stop her. A chipmunk started chattering at us as we passed the tree it clung to, indignant as it skirted backward and up to the lowest branch.

"I just can't believe our home will be gone. I always thought we'd figure out a way to keep it."

For months, the idea of having to say goodbye to the home she grew up in was harder to comprehend than the thought of her mother passing away. I understood her distress, felt it too. But I couldn't figure out why it was more difficult for us to say goodbye to a landscape than to a human.

My grandmother was a living, breathing, beautiful piece of life, a piece of the world we had grown up knowing. She had created my mother, and therefore, me, inside of her, taught us to sing and speak and love. The house was just a structure of brick and wood and mud. My grandfather used to say, "Never shed tears over material objects, only shed them for a life." Yet even as our footsteps padded softly on the leaves underneath us, I found the corners of my eyes stinging, not for the first time, when thinking about my grandmother's home.

Why should we grow so sad at the thought of a building being demolished?

We invited Lynnie and Rich over for dinner and sat on the patio with them, drinking some of the remaining wine from my grandfather's cellar. They talked about the new house they will build here. Lynnie declared it would be made purely out of cement, glass, and marble. She admired the massive and ancient oak that still stood outside my grandmother's window—despite that it died from a disease that poisons many of the oaks in the area—and compared it to a sculpture. Rich announced that they hired a famous architect to design the house, and told a story about a shower he built. A shower that began as a solid block of marble in Italy, then was shipped to China to be cut, shipped back to Italy to be molded, then shipped to the West Coast of America. It was then picked up by helicopter and lowered directly into the bathroom of a vacation home in Pebble Beach.

Lynnie said, "What I love about this place is it's so quiet. Isn't it so quiet here?" and we all fell silent, but I didn't hear any silence. From behind came the *caup* of the two ravens that had been lounging on various trees around the yard for the past hour, and the *whoosh* of air through their opal and ink feathers as they'd take flight from one oak to the next. Miguel, my grand-mother'sgardener, was rummaging and moving tools around the corner of the house. Seeping out of the landscape around us was the high buzzing of crickets and the distant bark of a dog and the rustles of leaves as a lizard scurried to cover and the wearied sigh of a mourning dove from one of the orange trees. On a different day or hour there might have been the cooing of wild turkeys, or the yips of coyotes during the night. I could practically feel the vibrations from the agave and jade around us breathing, taking in carbon dioxide through their stomata and releasing oxygen. From somewhere in the house my grandmother was singing, singing to me from a distant memory.

"See? So quiet," Lynnie said.

While we sat on the patio discussing sculptures and showers, the air around us was hazy and pink and smelled of smoke. One of the worst wildfires in US history was running rampant north of us, burning the land, destroying homes, and taking lives in Sonoma County. We could feel the dryness in the landscape and in our bones. The fire that started north of San Francisco could just as easily have started here.

Many of the dead were so burned that the only way of iden-tifying them was through mechanical body parts—artificial knees and pacemakers. Many who perished were older, unable to escape their homes quick enough. If the fire had originated in Portola Valley, and a few months or years earlier, my grand-mother and her home would have been reduced to ashes. She'd been confined to a wheelchair and then a bed during the last few years of her life; they would've had to identify her bones by her pacemaker and titanium knees.

The fire came after a heavy winter; many believed the infamous California drought gone after the large amounts of snowfall in the Sierras. Yet the fire proved to be one of the most devastating in American history. It's a misconception that a wet winter diminishes the threats of fire in the summer; if the following summer months experience a drought then a wet winter increases the risk of wildfires.

Winter sets the stage for the growing season during the rest of the year. A heavy winter replenishes watersheds and encourages vegetation growth, but if a drought follows the influx of moisture, all the new vegetation will dry out. In Northern California, the heavy winter of 2017 led to a rapid and wild growth of underbrush. The following summer drought baked that underbrush into kindling. Fire spreads much quicker and much more violently through cramped underbrush than it does through a canopy of trees. After six months of drought, the oak-strewn hills were ready to burst into flame at the merest spark.

The winds of the West Desert are simultaneously great for the Wasatch snow and destructive to the California coastline. They dry up the storms coming in from the Pacific and contribute to the soft, fluffy snow that falls in Utah. But when the winds circulate back to the coast, they bring the dry desert heat with them. Known as the *Diablo winds* in the north and the *Santa Ana winds* in the south, they are most common during autumn, prime fire season. When there is a low-pressure system off the coast of California, it draws the air from the high-pressure system over the Great Basin toward the low-pressure system over the ocean. These katabatic winds pick up momentum and heat as they move from high pressure to low pressure, blowing hot and fast. The wind responsible for the fire clocked in at eighty miles per hour and ninety-one degrees Fahrenheit. Though these winds are natural, they are predicted to become more violent as climate change alters the jet stream.

Fires used to be more frequent in California, likely due to burning practices by Native Americans. Preferring the open, arid lands for traveling and hunting purposes, they'd set fire to the landscape, creating more frequent and smaller fires. Consistent burning created what is known as California's *mosaic ecosystem*, and without the practice the land becomes more susceptible to destructive fires. When we combine a landscape prone to burning, a wet winter and a dry summer, intense winds made worse by climate change, and sprawling infrastructure that borders wilderness—wildland-urban interface areas—we increase the risk of extreme wildfires.

While the wildfires still burned we scattered my grandmother's ashes in the hills above her house. Loading everyone into cars, we drove and then hiked up to the place she had requested.

Everyone was singing. The noise was oddly muted, the moss and spongy soil absorbing the sound rather than echoing it around the forest glade. I stood at the base of a massive Douglas fir, my open-toed hiking sandals sinking into the detritus of the steep hillside. The fir's girth was hard to comprehend; each gnarled branch stretching out of its monster trunk could have been a separate tree. Indeed, its branches were thicker than most of the surrounding trees. Its odd, towering shape was achingly familiar, one I had visited countless times with my mother. This fir is where my family scattered my grandfather's ashes twenty-five years ago. Whenever we came into town my mother would set aside time for us to hike up to the tree to visit him. I never truly knew him, since I was a toddler when he left, but I know the tree that his ashes fed, and I know the hike and the hillside and the landscape that his body returned to. It's where his soul lingers—his soul's place. We were reuniting my grandparents in the hills they loved together, where they could watch over the adobe house they built in the valley below.

I moved the guitar around my neck so my sister could pass me the bag of ashes. Taking the dust in my hands, I thought of my grandmother's laughter, of the way her eyes would light up whenever I'd bring a vase of freshly cut flowers from her yard to place beside her bed. The substance was coarser than I expected. Breathing in the smoke of thousands of oaks and homes burning north of us, I swung my arm in an arc, the motion of a brush-stroke on canvas. Her ashes created a constellation on the rough, dark bark.

Turning from the tree, I thought about my family, my mother, and how, if another fire started here and this tree burned down, we wouldn't have a place to grieve.

Ashes to ashes to ashes.

Just over a year later, another even more devastating fire tore through Paradise, California. The Camp Fire killed almost ninety people and decimated an entire town, the Diablo winds pushing the fire faster than residents could evacuate. At the same time, another fire destroyed one of the nation's wealthiest communities in Malibu. And yet the phrase "climate change" was rarely used in the reports, despite four of the worst wildfires in our country's history happening almost within a year of each other. Despite the fires following two of the worst summers for wildfires in the West.

As stories of the survivors began appearing in the headlines, as well as stories of those who perished, I grappled with the possibility that someone I loved, or myself, could be killed by a wildfire, or another side effect of climate change. I imagined trying to evacuate my children from a burning home. The fear, the loss, even though hypothetical, was too great to confront.

The taxi arrived too soon. Much too soon, despite the fact that we were already running late for our flight. I was standing on the patio outside my grandparents' house, the afternoon after the celebration of life. Everyone else was leaving the next day, but I was returning to Utah that night with Colin.

The car's muted engine rumbled from the driveway. I hadn't said goodbye to anyone yet. Earlier I had walked through the house, touching the splintering wood panels, listening to the echoes the slap of my sandals made on the cold hallway tiles, trying to grab hold of something to help me through this. The motions reminded me of when I had to say goodbye to my childhood dog, Sasha. I held her one last time, trying to take the warmth of her body into mine, to keep the memory of her inside me forever. Though I tried to draw out that moment of goodbye as long as I could, the actual process of letting her go occurred too fast, too violently. When I let go of her, rather than taking part of her with me, it felt like a part of me had been ripped out.

This landscape was slipping through my fingers like dust. I tried to capture the details of the distant oak hills and the scent of citrus breezes from the orange trees. The land felt like it was crumbling beneath me; I was sure I must be standing on quicksand. Some primal part of me felt the urge to run away, to disappear into the hills so I never had to say goodbye. My eyes filled with tears faster than I could wipe them away, blurring what I knew to be my last vision of this place.

Memories were sewn into the landscape, everywhere I looked. By the large oak I saw my six-year-old self during my aunt's wedding, wearing a purple-and-white flower-girl dress and a big white sun hat, my father lifting me onto the lowest branch despite protests from my mother. On the small patch of grass around the side of the house, my grandfather was hitting golf balls into the valley below. My own children giggled from beneath the orange trees, their blue eyes peering up at me through the green foliage—memories that would never happen.

Footsteps scuffed the ground behind me: Colin walking out of the breezeway. I hoped he wouldn't walk up next to me. I knew I couldn't hide my tears. I could sense his hesitation; he didn't know how to console me.

If I could have spoken, I might have been able to tell him that just standing there, existing in the landscape with me in that moment, was enough. But my voice had gone.

Colin cleared his throat softly, a slight sense of urgency in the sound. We needed to leave. We should've left ten minutes ago. And I still hadn't said goodbye to anyone.

I turned away from the yard. Averting my eyes so he couldn't see how red they were, or that my makeup was running, I brushed past him, trying to swallow a sob. I stood in the dark breezeway for a moment to recapture my wits. The closeness of the adobe walls, the way that they held the cold air between them even on the hottest of days, was reassuring and calmed me a little. I opened the door to the kitchen.

For the last three generations, only women have been born into our family. My grandparents had three daughters, then eight granddaughters, and now three great-granddaughters. And be-sides two of the great-granddaughters, we all have eyes that are varying shades of blue, from my sister, whose eyes resemble the light, icy blue of a winter snowpack, to my mother, whose eyes contain the turquoise depths of tidal pools.

I opened the door, and there was the anchor that I'd been desperately trying to find in that moment of goodbye. Ten sets of blue eyes looked up at me. My six cousins, my sister, my aunts, and my mother all stood in the kitchen, peeling off hundreds of pictures that had been taped to the pantry doors for decades, tossing them into cardboard boxes that would be packed up and brought to our rooms at Snowbird. They had hair the color of golden wheat to dark chestnut to strawberry blond; some had olive skin while others had freckles, but all had blue eyes.

Each woman approached me to give me a hug, and I knew why leaving this place was so hard. The house was an ecosystem, made of both living and nonliving things, of love and memo-ries. We weren't just saying goodbye to a physical thing, we were saying goodbye to a piece of our family, a piece of our identity: one of the landscapes that held us together.

I left the breezeway and walked onto the driveway. Colin was waiting next to the taxi, and I finally mustered the courage to meet his gaze. His eyes were the lightest, most arid shade of jade, the meeting of earth and atmosphere, so crystalline they could blend into the sky around them.

III. Frosting Season

Bonding

It was a November day, crisp and clean. My sister and I opened the door to my mother's Suburban, shouldering our comically large backpacks. We could feel a certain scent in the air. Walking through our elementary school's campus, our bright black shoes scuffing the layer of frost that had crusted on the grass overnight, we made bets on when there'd be a layer of snow on the ground. I predicted by the time we left school. My sister guessed by recess.

Tyndall and I were professional snow predictors, but not because we knew how to read weather radar maps, like our father could. We could smell the snow before we could see it. Snow technically has no real scent, but there was no denying that we could *sense* its scent.

Rain, on the other hand, does have a smell. Upon falling to the ground, it mixes up oils that have collected on rocks, plants, and soil, releasing these oils into the air. Rain smells like earth, which makes me tempted to claim that snow smells like the sky. But rain can *actually* smell like the sky, or more specifically, ozone. During thunderstorms, energy released by lightning combines oxygen atoms (which are just one single atom—O) with oxygen gas (a more stable form, where two oxygen atoms are joined—O_2), resulting in O_3, or ozone. Ozone comes from the Greek word *ozein*, which means "to smell," and has a defini-tivescent to it—one that we register as airy, pungent, and clean, since it smells similar to chlorine. During a thunderstorm, wind can blow ozone in front of the storm, and as a result we taste the storm on the horizon.

There is no scientific proof that snow smells like anything. Yet it undeniably does—it smells bright and cold and clear. It smells crisp and soft at the same time, distant and close. It's a delicate scent, yet unattainable; the odor of twilight, the aroma of twinkling light. If rain smells like the earth and lightning like the sky, then perhaps snow smells like starlight.

Once the snow started falling properly, we'd participate in our favorite sport of the season: snowflake hunting. It didn't matter where we were: at school with our friends or on chair-lifts with our cousins. We'd stick our gloves out, sometimes palm up, sometimes palm down, and watch as the tiny white crystals decorated our outstretched hands. We had figured out that the darker the clothing was, the better we could see the white shapes. We were looking for the most perfect snowflake.

Sometimes the flakes were so small we'd have to get our pink-tipped noses a few inches away to pick out their patterns. These seemed too delicate to exist. Other times, fatter flakes would stick together as they fell, settling on our jackets in clumps of four or five. No matter the size or shape, we studied them intensely as they landed, trying to find the most symmetrical, the most intricate, the most beautiful. Sometimes we'd come across one that seemed like it had fallen from our imaginations, looking too mystical to be from this world. No matter how unique the crystals, we'd only be able to appreciate them for a few seconds before their six tips would begin to fade, disappearing into the material that held them.

In 1611, mathematician Johannes Kepler became enamored with snowflakes. After watching them fall on his coat sleeve one night, he realized that, though each individual crystal was unique, they all had six points. Using what he knew about shapes of the natural world, he tried to solve the mystery of snowflakes using mathematics. He compared the six corners of snowflakes to hexagonal honeycombs. Of all the geometric shapes in the world, only three lay completely flat against each other when arranged in large numbers: the triangle, the square, and the hexagon, and of these the hexagon creates the most compact design, which is why honeybees create their hives out of them. But Kepler couldn't figure out what compactness had to do with snowflake shape. He never discovered why six was the magic number for snow.

For centuries, philosophers, cultures, and scientists have assigned numbers to certain parts of the natural world:

Zero: emptiness, nothingness, void
One: the creator, or creation
Two: the celestial bodies in the skies: the moon and the sun
Three: the base and flames of a fire
Four: the stretching corners of the sky, accompanied by the four strong winds
Five: the hands of life, found in flower petals, leaves, and our own bodies
Six: water, as liquid, vapor, or frozen into a six-pointed star

Humans have known that six is somehow the number that defines water, though no one knew quite why until the start of the twentieth century, when microscopic chemists began discovering the structures of molecules. Scientists realized that Kepler's unsolved problem wasn't a problem of mathematics; it was a problem of attraction and repulsion. Life exists on earth and snowflakes have six points because of hydrogen bonds.

Water molecules are composed of two hydrogen atoms and a single oxygen atom. Oxygen atoms are composed of a nucleus, which lies at the center and is made up of eight protons and eight neutrons, and eight electrons that orbit around the nucleus. Hydrogen atoms have a nucleus made of just one neutron and one proton, with one orbiting electron. Atoms bond together by sharing *valence electrons*, or electrons that are "free" and not in a pair. This process is called *covalent bonding*. For a hydrogen atom, this means borrowing one electron so its single electron becomes a pair. Oxygen atoms have two pairs of electrons and two valence electrons. They crave two more electrons to make them "complete."

When hydrogen and oxygen are both present, two hydrogen atoms will line up next to an oxygen atom so that one of the oxygen atom's free electrons can join each of the hydrogen atoms, giving the oxygen atom the two electrons it wants, and each hydrogen atom getting an electron in return. Because the electron pairs of the oxygen atom and the hydrogen atoms all repel each other, it creates a tetrahedral shape, with oxygen in the middle and the two hydrogen atoms and two pairs of electrons as four legs sticking out of it. Because the four electrons—already paired up in an oxygen atom—have no purpose when it comes to joining molecules, they are typically left out of diagrams. Without them, the shape of two hydrogen atoms with an oxygen atom in the middle looks like a widened right angle, an angle of 105 degrees. The resulting molecule is H_2O, or a molecule of water.

Oxygen holds the electrons closer to it than the hydrogen atom does, which creates a polarity—the oxygen negative, the hydrogen positive. When two or more H_2O molecules come close, the negative oxygen of one molecule will attract one or more of the positive hydrogens of other molecules. The bond that forms between them, positive and negative attracting each other, is called a *hydrogen bond*. This bond is what creates the shape of water.

Hydrogen bonds are strong enough to hold molecules of water together until *thermal energy*, or heat, breaks the bonds, separating the H_2O molecules from each other. Individual molecules take the form of gas rather than liquid, which is what happens when water is boiled. However, hydrogen bonds overall are weak, and atoms and molecules are erratic. When in liquid form, these molecules jump and jitter around, constantly moving, breaking and forming new bonds with neighboring molecules and atoms. As the temperature cools they slow down, breaking and forming at slower intervals until they stop altogether, hitting water's freezing point.

Because of the tetrahedral shape, the hydrogen bonds that connect separate water molecules together continue to connect them at 105 degrees, creating a hexagon with six sides and six corners, an H_2O molecule at each corner, held together by hydrogen bonds. This shape is what water takes when it changes from liquid form to ice, known as *crystallization*. The hexagon holds more space within its shape than when the atoms and molecules are moving around, which is why water expands as it freezes into ice, and why ice floats on top of water.

This hexagon is also the reason why every single snowflake, no matter its uniqueness or size, has six points: the answer to Kepler's unsolved problem. A hexagonal plate is what every snowflake starts out as. When hexagons continue to build and attach to the original, they spread out horizontally from the six corners of the hexagon into six arms, as well as vertically, creating what's known as a *hexagonal lattice*.

For this process to create a snowflake, water molecules need some sort of particle to latch onto, a sort of nucleus or center to start the process to make individual snow crystals. This particle is usually a piece of dust or pollen floating in the atmosphere. The conditions present at the time the crystal begins forming determine how each snowflake will look. Temperatures, humidity, and wind currents influence the shapes of snowflakes as they fall through the air, gathering more moisture and water molecules, or losing them in the twisting winds.

Eventually, when they touch the ground, or land on other flakes already laid to rest, they can change again. Because the earth is typically warmer than the air during the winter months, most snowflakes will melt if they land on bare soil. But once a sort of momentum picks up, where more snow is falling and landing faster than the snow before it can melt, it begins the building process that leads to a layer of snow, and the beginnings of a snowpack.

Immobilizing

There was no snow. Not in the valley, not on the benches, not in the mountains. It was only ten days away from Snowbird's intended opening date and less than twenty-four hours after Donald Trump was elected forty-fifth president of the United States of America. A blanket of white covered the city beneath me, but it wasn't snow. It was Salt Lake City's infamous inversion, which came two months early.

I'd driven up a hill behind my apartment, just north of Salt Lake City, to try to gain perspective—on what, snow? the election? the inversion? I wasn't sure—and had been sitting in silence for half an hour. My eyes stung. From my feet the hill dropped down, leveling out into the submerged city, and from there the grid system sprawled south and out of sight into the white. I could barely see the Oquirrhs across the valley, the outline of their ridges murky shadows in the haze. The Wasatch were more defined due to their higher elevations, but their bare peaks were a sad reminder of what was missing that November day. Buildings and smokestacks alike rose eerily out of the murk, looking post-apocalyptic. Cars moved on the freeway, their metal hoods reflecting the dull sun. They looked like little electrical currents moving through a wire.

The Salt Lake Valley was supposed to be well on its way to winter, but I was wearing a skirt and sandals. The air filling my lungs was toxic enough to merit wearing a face mask. And the night before, the United States of America elected a man into office who doesn't believe in climate change.

I woke up that morning feeling numb. My apartment was a disaster. Plants had been knocked over, the floor was covered in dirt, the countertop was sticky with spilled drinks, and my laptop, which we had used to watch the election, was broken. I walked through the apartment as if in a dream, feeling disconnected from my body.

Did our party really get this out of hand? I wondered, stepping over broken glass in the kitchen. *Is this really the reality I'm waking up to? A Trump presidency?*

My roommate's door was closed. I knocked on it softly, pushing it open when she responded. Kailey was curled up in bed, crying silently into her pillow. I sat down on the side of her bed, placing a hand on her shoulder.

"Are we going to be okay?" she whispered.

I had no words to answer her.

Emotions evaded me the rest of the day. I knew I had to be in shock. That evening I drove up to the empty lot above my apartment to try to get my body to feel something. Anything.

To my eyes, hate and ignorance in the form of smog infiltrated everything, weaving through trees and buildings alike, staining the sky and muffling the sun. An uncomfortable buzzing had started in my head, and I knew it was from the air: 2.5 particulate matter creeping behind my stinging eyeballs and navigating around the soft hairs in my nostrils to make its way into my brain and bloodstream.

I knew the snow would come. Maybe not in the next week, but eventually it would come and lay itself across the body of the Wasatch like a lover long awaited. But how long would it stay? It could continue to come back year after year, or it could flicker and fade like a dying lightbulb, finally calling it quits.

Last week Kailey and I drove around this neighborhood, looking for wild succulents. When we'd find some we'd drop to our hands and knees, gently digging them out of the ground so we could replant them in pots in our apartment. There were a few growing out of a pile of rocks by my feet, sage-green and soft. A few days ago I would have stroked their leaves, marveling at the way they could grow with almost no soil. Now they just reminded me of the plants still lying on my living room floor. I had not yet found the energy to pick them up.

I closed my eyes. Two birds were flitting in a tree somewhere to my left, but I had no desire to identify them. What I would do to turn into a bird that very moment and fly away from this place and pollution, to be ignorant of climate change. I would trade my lips for a beak and my voice for a song. I could spend the rest of my life singing and never have to speak again of pain and suffering, or convince myself or anyone else that everything was going to be all right.

A man called from somewhere below me. A boy responded. There was a dull thud, like a foot kicking a ball, and the boy laughed, his joy cutting sharply through the haze.

My throat constricted. For the first time that day it seemed, my heart started beating, rapping on the inside of my sternum like knuckles. A boiling sensation started in the deep space beneath my stomach. I suddenly felt nauseous. I rubbed the inside of my stinging eyes, and my hands shook as I raised them to my face.

As the election results flashed across my laptop last night and the wine in the bottle in front of me ran out, I had made a promise to myself. In the face of climate catastrophe and un-curbed carbon emissions, I knew the most significant choice I could make as an individual would be to refrain from starting a family.

The sun dangled closer to the Oquirrhs, casting a putrid light that turned the white haze yellow. As the light shifted the freeway receded into shadow and something small was lit up, waving its red, white, and blue into the pollution. It was easy for me to feel hate and dispassion right then. The majority of Utahns had voted for dirtier air, more-congested freeways, higher asthma rates in children, and less snow. And America? We voted for sexual assault, white supremacy, and climate catastrophe. The snowless, poisoned valley I was staring into was our future.

The boy and his father laughed again. I blinked; a tear slid down my cheek.

I will never be a mother.

Reflecting

The heels of my boots clicked against the concrete in the hallway, echoing around the narrow space as I approached the door that would lead me out to the parking lot. I stopped at the exit, taking a moment to find my bubble-gum-pink Taser in my purse before I left the protection of the building.

About a month ago a woman was raped in her car in a nearby parking lot, in the middle of the day. Since the assault, I had made it a habit to carry my Taser with me after my night class, dreading the short but dark path from the building to the parking lot. Though I appreciate the green and golden colors of the day, at night they disappear in the vacuuming darkness between stars.

Gripping the Taser firmly in one hand, my other hand found the cool metal of the door handle and pushed it open to . . . light.

The sun had set hours ago, yet the air around me glittered, as frosted and crisp as cracked ice. The sky and ground alike, the buildings, discarded pine cones, cars, and trees all seemed to glow with a soft white light from within. The world had been bare as I drove to class; now a few inches of snow covered everything in sight.

My fingers released the Taser as I crossed the parking lot. The woman was attacked in daylight, yet I felt more comfortable crossing this place alone by snowlight than by sunlight. Sunlight, coming from a single direction, ignores what it doesn't touch, effectively neglecting half of the world at any given moment. Snow, on the other hand, borrows light from other sources. When there is light present, even indirect light, snow diffuses it between its six points before reflecting it outward, radiating from within.

When light waves (or particles, since light is both a particle and a wave) travel through a substance, some of the waves are absorbed while others bounce off other particles. Electrons have *frequencies*—the speeds at which they vibrate—and if the frequency of the light wave is the same as the electron it encounters, the electron absorbs the vibration of the light wave, converting it to thermal energy. If the frequencies of the light wave and electron are different, the wave will be reflected off the atoms, or scattered. Longer waves, colors closer to red on the color spectrum, are absorbed while shorter waves, blue tones, get scattered. This is known as the *Rayleigh Effect* when the particles doing the scattering are smaller than the wavelengths of light, and as the *Tyndall Effect*—named after a nineteenth-century physicist named John Tyndall, not after my sister—when the particles doing the scattering are much larger than light wavelengths. This is why the sky appears blue; blue light waves scatter when they collide with particles in the earth's atmosphere. It's also the reason water appears clear in small quantities; water electrons don't vibrate at the same frequency as light waves. The more dense the water is, the more red light is absorbed and blue light is scattered—this is why the ocean is cerulean to our eyes. This is why snow can appear blue sometimes, like in a deep footprint or a glacier crevasse.

Most light waves that move through snow are reflected, making the snow appear white: the color of sunlight. Freshly fallen snow, like the snow I experienced while walking to my car, reflects and diffuses light especially well, before dust and other particles able to absorb light waves integrate into the layers. This reflective quality, called *albedo*, is due to the shape of snow crystals and how they form.

A single snow crystal contains flat, crystalline surfaces, known as *facets*. A facet is like a geometric plane; the flat surface of a certain shape. The word comes from the French word *faccete*, meaning "little face." Minerals develop facets, and humans cut facets on to diamonds and gems to make them sparkle when they catch light. Facets are responsible for snow's reflective properties, and are why freshly fallen snow is able to diffuse light so well.

When water molecules begin latching onto each other, forming hexagonal plates, free molecules are drawn to the uneven places in the shapes—discrepancies where there is more molecular binding potential. They are less likely to bond to an already flat surface of water molecules. Because of this, they create a smooth hexagon shape—a facet. Facets will continue to build on each other in two different forms: basal facets, which are flat, and prism facets, which build a more three-dimensional shape. To create the most simple snow crystal, two basal facets and six prism facets are required, forming a hexagonal prism. As environmental conditions influence the development of a single snow crystal, branches and ridges and rims, all made up of facets, form and build on each other, causing the snow crystal to expand and become more intricate.

Each facet provides a surface that light can be reflected off of, so with the most simple snow crystal, with two basal and six prism facets, light will be diffused from as many as eight flat surfaces. *Diffusion* is when light from a single source gets scattered in many different directions rather than just reflecting in one direction, like if a beam of light were to be reflected off a disco ball rather than a flat mirror. This, ultimately, is why snow sparkles; each snowflake is a little diamond, reflecting light from a multitude of little faces.

Snow reflects brightly when under direct sunlight, and it seems to glow from within when there is indirect light, even after the sun has set. This effect is even stronger when the sky is overcast, effectively trapping all light between land and sky as it did that night as I crossed the parking lot.

Snow plays an important role when it comes to regulating the earth's temperature. The North and South Poles reflect 80–90 percent of light that reaches them, rather than absorbing it like the darker oceans, rocks, soil, and flora do, which have less albedo than snow. Bare ground absorbs four to six times more of the sun's energy than snow does. Snow helps to keep the globe cooler than if the equivalent of the reflected light were getting absorbed and converted into heat. As climate change causes the snow and ice at the poles to melt, less light will be reflected back into the atmosphere and more heat will be absorbed, further contributing to the warming of our planet. Though the poles are the largest areas that do this, snow all over the world contributes to this reflective process. Snow in the Northern Hemisphere is especially important in regulating the earth's temperature, since 98 percent of seasonal snowfall occurs north of the Equator.

Low-snow years will lead to a warmer climate, and a warmer climate will lead to more low-snow years. It's the most vicious of cycles—a positive feedback system. The more the glaciers continue to calve in places like Antarctica and Greenland, the less light will be reflected, and the warmer the oceans and earth will continue to get. Though the poles reflect the most sunlight because they stay white no matter the season, if we lose our snow in the American West it will have global repercussions.

Albedo is the reason I'd stay up late during the winter as a child. A peculiar thing occurs when there is no direct source of light during the winter months and when snow is on the ground. After a fresh snowfall, and once the sun has set behind the

Oquirrhs, the Wasatch can glisten with a pale, mauve light. The Germans have the best word for this phenomenon: *alpenglühen*— or "alpenglow" in English. Alpenglow is caused by light from the sun bending around the earth to light up mountains indirectly, which is emphasized when snow is present to diffuse the light over the entire landscape.

In order to witness this glowing world I'd pull my blankets up to my nose and keep my curtains drawn back, though it meant that my room would drop significantly in temperature. Colors might drain under the cold twinkle of the stars, but snow shimmers even brighter.

Crystallizing

One morning, on the cusp of winter, Colin and I stay in bed for as long as possible. In the window above our heads are the tiniest of snowflakes. Too delicate to fall to the ground, they float in the predawn air like stars in the sky. Diamond dust.

Something had shifted between us in the last few weeks. There were moments when words would catch on the tip of my tongue, and I would only just be able to hold them in before they slipped out and exposed me.

Even then, lying with him, something threatens to break free from me. I hold it back, convincing myself that I might frighten him away.

"I love you," Colin murmurs.

The world has frozen over that morning. Fallen leaves are curled and frosted, their edges and veins stenciled with delicate designs of white. The lawn looks like someone has scattered hundreds of thousands of little diamonds across it. On the windshield of my car are constellations of crystalized stars, sparkling golden as the morning sun shines through their tiny, outstretched fingers. Winter is on its way.

I drive up Little Cottonwood Canyon after leaving Colin's house, and am soon jogging through a landscape ripping along the seam between seasons. Snow, lingering from the last storm, outlines the shadows of thick pine boughs. Where the sun can reach, glistening drops of silver, water molecules turning from solid back into liquid, dangle like earrings from the drooping tips of grasses.

The soles of my shoes pad softly across fallen aspen leaves as I make my way slowly uphill. In a matter of minutes, the small specks of crystals hanging in the air become fat flurries, twirling their way to the ground in clumps. Quickly and quietly, the world turns to a blizzard. Aside from my laboring breaths, it is oddly silent. Perhaps most of the local animals have already burrowed into hibernation, sensing the long nights ahead. Strange, I think, how some animals sleep the winter away while others come alive. Namely the ones who strap long pieces of plastic and carbon on their feet so they can go to the top of a mountain in a storm. Surviving a winter takes effort, even from a human's perspective, but these frozen mornings of silence are worth it. I certainly can't complain; I have the entire mountain to myself.

I round a corner and come nose to nose with a bull moose. Or that's what it feels like. I stop abruptly, my breath catching in my lungs. It is over a hundred feet away, blurred by the ribbons of white weaving through the air, though the shock makes me feel like I have run headfirst into it. It has either sensed me coming or just happened to be looking in my direction, for its dark eyes hold mine from across the frost-streaked field.

I wonder if I am safe to appreciate the moment. A few years ago my dad had been charged by a bull moose in our backyard when the sound of his camera phone clicked in the silence between them. The moose had stopped a few feet from him, but it could have easily gouged my father if it had found him a real threat. I know that, even at this distance, I won't have much of a chance to escape if it decides to charge.

Well, I won't give it reason to charge, I think bravely, but then I notice the female just uphill of it. And then a calf, grazing half-way between the bull and I.

The bull snorts, blowing cold smoke from its flared nostrils. With lumbering steps it begins moving toward me, scuffing the fresh layer of snow with its heavy hooves.

I back up, scattering the yellow leaves, and disappear around the corner, breaking into a full run once I'm out of sight.

How silly of me to think I was the only creature appreciating the silence of the morning.

Condensating

One Thanksgiving, in 2001, there was no snow. Not even a dusting. This was unusual; sometimes on Thanksgiving we'd be treated to a few inches of powder on top of early season conditions. My family was staying up at Snowbird and even in the mountains there wasn't enough snow to make a snowball. Then a massive storm rolled in like a tidal wave, closing the canyon roads for half a week and forcing Snowbird to declare "interlodge"—when guests and employees are legally required to remain inside the lodges due to avalanche danger—for two straight days. It was deemed the "One Hundred Inches in One Hundred Hours" storm, though in reality it was two storms that came one right after the other, each contributing approximately fifty inches of snow.

When a storm cycle passes over a landscape it can merge with a variety of atmospheric conditions, which in turn can alter the way the cycle moves, the amount of moisture present, and the atmospheric pressure of the surrounding areas. Atmospheric science is complicated and can be tricky to comprehend, but I've done my best to synopsize what occurred during the two storms.

The first storm system came from the Pacific Northwest. A cold cyclone formed near the southwestern coast of Washington, weakening as it hit land and moving south, overtaking a warm front moving north somewhere over Nevada. When this occurs, the heavier cold air sinks underneath the warmer layer, often creating a tongue of warm air that will hit an area just before the cold front does. This is why the Wasatch is often hit with pleasantly warm weather right before a storm system moves in. If the air in front of the warm front is colder than the cool air overtaking the warm front, it's called a *warm occlusion*. Warm occlusions often lead to higher rates of precipitation because the approaching cold front is not cold enough to sink under and lift the air that was already present. It was a warm occlusion that occurred during the first storm that hit the Wasatch that Thanksgiving, made more intense by what is known as *lake effect*. Due to the Great Salt Lake's relative warmth and inability to freeze from its salinity, cold northwestern winds often blow warm moisture off the lake after a cold front, creating further precipitation and adding inches to storm totals.

The second storm system, moving in immediately after the first dissipated, originated in California. This front became more intense as it crossed the higher elevations of the Sierra Nevada, merging with a weather pattern known as a *lee trough*. This phenomenon contributes warmer and drier air to the developing system and occurs in the dry deserts found on the lee sides of mountain ranges. Once this system crossed the Oquirrhs, hitting another, smaller lee trough, moisture from the Great Salt Lake added to the intensity of the storm and the extreme levels of precipitation. This is a classic storm pattern that winds up in the Wasatch; originating in the Pacific, dropping massive amounts of wet snow over the Sierra Nevada, drying up while crossing the deserts of the Basin and Range, and dumping again over the eleven-thousand-foot peaks of the Wasatch, often gaining more momentum from the Great Salt Lake.

Snowbird reported eight feet of snow, Alta ten, and Tyndall and I missed two days of school. Not a single car in the parking lot could still be seen under the snow, just slight bumps indicating where a car was buried—the only way to tell that the parking lot wasn't just a flat plot of empty land. Once it finally stopped snowing it became an enormous task for hotel staff and guests to dig the vehicles out. The Lodge lost heat and electricity, so my mother and grandparents placed candles from our storage unit around the room and pulled out all the blankets and pillows so Tyndall and I could build a fort. To cap off the enormous snow week, Tyndall and I created and performed an intricate, hour-long puppet show with flashlights for anyone who would watch. My mother and grandparents spent the entire time moodily staring out the window at the snow-covered hills, frustrated at being unable to appreciate the fresh snowfall.

Bristling

In 2017, Snowbird opened its lifts to the public and closed them again within three days. Enough snow had fallen over the past few weeks to fill in the high-elevation, north-facing slopes, but the majority of Snowbird's terrain was still barren and brown at the end of November, and temperatures were too warm for the resort to successfully make man-made snow.

Yet Snowbird still opened.

I visited the resort a week before its intended opening and was shocked at how little snow there was. From Snowbird's Instagram account, which featured photo after photo of powder-skiing shots, the public impression was that there was enough snow to get the winter show on the road. Myself included—until I drove up the canyon and saw that Snowbird was still in mud season, not even close to winter.

I didn't think there was any way that they would be able to open in time, and no possible way that they should try. But they did. Months before, Snowbird sold package deals for the holiday: hotel rooms at one of the lodges, lift tickets, and Thanksgiving dinner grouped together in one inclusive price. From a management perspective, Snowbird *needed* to open for Thanksgiving.

The holidays are a very important time for ski resorts. In order to support their operational costs, many rely on those busy times to stimulate their flow of income. For some, the ability to open prior to and do well during the holidays is critical to keep the resort open for the rest of the season. But with a shifting climate that's been pushing the start of the snow season back, operating during the holidays is becoming more and more unreliable.

The ski industry survives on a few key environmental factors. One is the terrain, and with it gravity; it requires a certain steepness to get skiers and snowboarders to travel down a hill. Another is elevation; typically, the higher the elevation, the colder the temperatures. Latitude and longitude, the distance a resort is from the equator and poles, affects its exposure to the sun during the winter. The way that a storm track moves across a landscape influences the local precipitation of an area.

With a shifting climate pushing the start of the season back, more resorts are relying on man-made snow to open in time for the holidays, making it possible for their businesses to exist. The beginnings of climate change in the mid-twentieth century coincided with the invention of the snow gun, a machine that creates man-made snow. As natural snow storms became less reliable, snowmaking became more accessible. According to the National Ski Areas Association, 89 percent of resorts in the United States use snowmaking to supplement their slopes when natural snow isn't enough. The European ski industry also has become reliant on snowmaking. Some resorts in Italy have runs with over 80 percent artificial snow.

Snowmaking is incredibly effective if the conditions are right, but depending on the humidity of the air, temperatures often must be at or below freezing in order for snow guns to work. If the air is very dry, snow can sometimes be made at temperatures above freezing. Snow guns run by being hooked up to some sort of water source, typically a river or a reservoir. If the water source runs dry, snow can't be made. At Snowbird, old mining tunnels have been blocked off with a plug to create an underground aquifer that holds about thirty-five million gallons of water. Some of this water is treated for drinking water, while another portion is used for snowmaking. It might be one of the more sustainable models out there; when the snowpack melts, it replenishes the reservoir, which doesn't lose moisture from evaporation.

Snowmaking is expensive, time consuming, and requires a lot of energy and water. Ski areas already have to dedicate anywhere from $500,000 to $3.5 million per year to run their snow guns. That cost, assuming that temperatures will stay low enough for resorts to make snow, will keep increasing as snowstorms become less reliable. Many resorts need 450 hours of snowmaking before they can open. But even if a resort does have the resources to make snow, if the temperatures don't cooperate even snow guns can't change an opening date.

While there are a few different types of guns used to create snow, most require a process that separates water molecules, turning them from liquid to gas. Cold air then blasts these molecules, creating a cold cloud. If the air is too warm, snow simply can't be made. Even if the process occurs, the newly made snow will just melt in above-freezing temperatures.

With the combination of less natural snow and higher temperatures, higher costs in snowmaking, and diminishing terrain leading to less incentive for consumers to continue to spend money at ski areas, maintaining a ski resort is already becoming less economically viable.

❄❄❄

It used to be a tradition for my father's side of the family to stay up at the Lodge over Thanksgiving, but after a few years without snow the tradition faded. In 2017, we decided to risk it and reserve our rooms for the holiday ahead of time.

In the days leading up to Thanksgiving, Snowbird hyped up the opening date by marketing it as "Game Day." In a gimmicky attempt to appeal to the masses, they created a logo specifically for the event and a computer-graphic video of Snowbird as a football arena that they posted on their social media pages. They managed to open one run with almost no snow, real or artificial. To carry the Game Day theme, they passed out foam fingers to customers waiting in the first lift line of the season. For the next week, these foam fingers scattered the bottom of the hill, soaking up mud.

A hoard of people on skis and snowboards cruised down a meager few inches of snow in fifty-degree weather. There were so many enthusiasts trying to take advantage of the limited terrain that one ski patrolman told my father it was the scariest day he'd ever worked at Snowbird. It destroyed the snow, and in three days the managers were forced to close the lifts again.

I chose not to ski on that opening day. It wasn't a noble decision on my part; I was experiencing lower back pain and didn't want to risk inflaming it. But my family skied, everyone from my cousin's toddler to my grandfather.

I sat alone in our rooms, staring out at the brownish hills. Sometimes, the business aspect of Snowbird, the side that uses marketing techniques to get more people on the mountain while largely ignoring the impacts of climate change, disheartens me. Skiing is a seasonal luxury, both from a consumer and a business standpoint. With climate chaos on the horizon, it's also a luxury that's not guaranteed in the future. Snowbird is making efforts to reduce their carbon footprint and move toward more renewable energies, but it's mostly behind closed doors. Promoting carpooling, providing charging stations for electric vehicles, cleaning up old mine tailings, and participating in a

"Slow the Flow" program, which promotes water conservation in the Wasatch, are all steps in the right direction. They are very important steps. Click on the "environment" tab of their website and you're presented with a long list of ways they claim they're helping the environment.

So why, I thought in frustration as I watched the single open lift spinning off in the distance, doesn't climate change come up in every conversation I have with strangers on chairlifts? Or even a fourth of the conversations? Or any of them, really? And why would Snowbird open without any snow, while posting staged photos of powder skiing on their Instagram, when instead they could have been promoting awareness of climate change, warming winters, and threats to the ski industry?

My family came back after a few hours, looking harassed, assuring me that I hadn't missed out on anything. But then my sister kicked off her boots and tugged the helmet off her head, pieces of hair coming loose from her long golden braid. She sat by me on the couch, pulling out her phone to show me the video she had taken of my father, my uncle, and my grandpa, skiing together in sync, and the picture of my cousin's two-year-old daughter, standing on a pair of skis for the first time between her great-grandfather's legs as he coaxed her forward.

Latching

In the fall we celebrated both Junior and Maxine's ninety-third birthdays. During the first few days of December, we celebrated their sixty-sixth wedding anniversary with dinner and a play, toasting to their marriage with glasses of burgundy wine.

After dinner, we walked outside and were greeted by the tiniest of snowflakes. How appropriate; the play that my parents bought tickets to was *White Christmas*. A few steps in front of me, my grandpa reached over and took my grandma's tiny hand in his own.

❄❄❄

After a fresh snowfall, snow crystals can latch onto each other like irregular puzzle pieces and settle on the smallest of surfaces, such as the thin white branches of aspen or cherry trees. Water molecules attract each other, a property called *cohesion*, allowing water to hold shape when other liquids can't. Water molecules also are attracted to other atoms, molecules, and objects through *adhesion*, giving them the appearance of embracing other materials, such as tree branches.

Stripped of their summer leaves, winter branches can become almost invisible to the naked eye until layers of snowflakes settle on their exposed limbs, adhesion and cohesion acting together. Then even the tiniest of gnarled branches become frosted with the soft crystals, impossibly delicate, each tree an intricate nest of white. Though trees are exceptionally hardy, even the most resilient can break under stress and circumstance.

The first child that my grandma delivered was dead. An unnamed girl, who my grandparents never laid eyes on, was strangled in my grandma's womb three weeks prior to her delivery date. When the doctor checked my grandma two weeks before her due date and couldn't find a heartbeat, he advised her to still go through natural childbirth. My grandma carried her dead baby girl in her belly for two more weeks before her water broke.

The Wasatch Front was slowly spiraling into the early winds of winter. The fruit trees in the family orchard had lost almost all of their leaves. My grandpa drove my grandma to the hospital and together they endured the twenty-four-hour storm of labor to deliver a figure of flesh and bone that would never take a breath of air.

They had a long winter. She laid her grief down between the cherry trees in the orchard and let the snow cover it. He sought deep, bodily breaths of the mountain air that his child never knew. Under the snow, the seeds of wildflowers that my grandparents loved so much—lupine, Indian paintbrush, columbine, and elephant's head—hibernated, protected from the winter frosts by the snow.

Then the melting season came, smearing the foothills of the Wasatch with mud. It reminded them that they shouldn't stay buried forever. That they couldn't. Spring was on its way and it couldn't be ignored. Melting turned to budding, and buds demanded attention. But the developing buds of their cherry trees, their largest crop, were never meant to open. A late frost caught them off guard during their most vulnerable time, freezing the swollen buds right as they were coming full term. Not one tree blossomed. Not a single soft petal could be seen among the rows and rows of dangling crimson fingertips.

IV. Season of Falling Snow

Turning

Skiing developed in the deep wintery hills of Scandinavia and the mountainous terrain of the Altai region of central Asia over ten thousand years ago because of necessity: an adaptation of humans to their environment. Populations living in areas where snow could cover the ground for over half the year required something to make traveling through the cold substance—not quite air, not quite liquid, not quite solid—possible, something with a surface area larger than human feet that would keep them on the top of a snowpack.

Early solutions included pieces of leather, light wooden blocks, and rope tied around wooden branches (a precursor to the snowshoe), all of which were strapped to the feet. The designs mimicked animals—like the snowshoe hare, lynx, and arctic fox—who had wide or long paws with thick fur to help spread their weight over the snow.

Skis were an upgrade from these inventions. The longer and thinner surface made skis the preferable way to travel; they were smoother, faster, and required less energy to use than blocks or snowshoes. Often the bottoms were lined with animal fur to create traction. Both women and men used skis for transportation and hunting purposes. By pushing themselves through the snow with a long stick, hunters could travel as fast as their prey. Deer and elk hooves often broke through the surface of snow or sank too deep to run fast, but humans had found a way to glide down a snowy hill.

Skis held such an important role in Scandinavian culture that they even had a place in Norse mythology. The giant Ullr and giantess Skaði were the deities of winter and the hunt, who traversed across the mountains on skis. In one story of Skaði she marries Njörðr, a god of the sea. They try to make their marriage work by splitting their time between the ocean and the mountains, but when each are away from their home they become heartsick; he for his fjords and islands, she for her mountains. Eventually, Skaði returns to the peaks to continue to ski and hunt, too attached to the landscape that shaped her to live elsewhere.

As society developed in these regions, skis became a vehicle to commute to work, visit neighboring farms, go to church, and travel to town to sell and buy goods. In the days when winter meant extreme isolation in rural populations, skis provided a means to stay connected to one's community. For women, especially, skis allowed the freedom to escape the remote confines of homesteads. Tax collectors and soldiers also utilized them, since skiing was similar to the speed of riding on a horse. Competitions involving skiing may have started out as military drills on skis, but they quickly became important events in Scandinavian society, and included cross-country races, downhill races, and ski jumping.

Then skiing moved south, becoming a thing of leisure. In Western Europe, the romantic era of the 1800s was shifting societal perception of the mountains. Landscapes once regarded as places full of darkness and death became places of adventure and desire, where one could experience the awe-inspiring might of nature. Small lodges and ski resorts began appearing throughout the Alps to accommodate wealthy adventurers.

Then Scandinavians began crossing the Atlantic to settle in America, bringing their skis with them.

In 1933, about half a century after the first skiers made their way to North America, Baqui strapped two barrel staves onto his feet at the age of eight. Using a stick, he pushed himself into motion and discovered what it was like to travel down a hillside on a primitive pair of skis, changing his course of life from farmer to skier.

Ice-skating was considered a mode of transportation back then. The farm that he grew up on in Provo had almost a mile of private road before it connected with the county road, so during the winter it was nearly a mile to the nearest plowed road. Once the snow on the farm road became packed down enough, my grandfather and his siblings used ice skates to glide across their family orchard. When the temperatures dropped below freezing, they would also use ice skates to travel up and down the frozen canal that ran through their farm. Occasionally, Junior's father, Levi, would flood a section of a field with water from that canal, allowing it to freeze across the field to create an ice-skating rink.

Barrel staves were supposed to be an upgrade from ice skates: another way to navigate across the farm during the winter. More surface area meant more distributed weight on the snow's surface. While ice skates only allowed for travel when the snow had been packed down hard enough or had become icy, the flat sides of barrel staves stayed on the surface of soft snow. And, most importantly for my grandfather, barrel staves allowed for downhill travel.

So, attaching leather straps to the staves, Junior began experimenting on a hill behind the family's barn. It was a rudimentary design at best, but my grandpa and his family hadn't seen the skis being produced in Europe at the time. He had never heard of skiing and didn't realize that he was mimicking a design that had originated over ten thousand years before.

Almost since he could walk, his body was developing the muscles required to push against the resistance of the ice while wearing skates, inconsistencies in the frozen water gripping the inconsistencies in the blade, and he grew accustomed to breaking free and coasting forward on one leg. Ankle muscles developed balance, core muscles developed stability, leg muscles developed momentum. These same muscles were the ones to help him transition from ice skates to barrel staves.

Barrel staves eventually turned to proper skis, bought from a sporting goods store in Provo by his mother, Jenny. As teenagers, he and a friend began skiing on logging roads up Provo Canyon. Others were starting to catch on to skiing, though mostly cross-country at that time. My grandfather remembers a hill that he and his friends would hike to, one that angled through a grove of aspen. It was a popular place for people with skis, but, as my grandpa tells it, no one back then knew how to turn while coasting downhill; there weren't any ski instructors in Utah. Those with a sense of balance could ski through the trees, leaning their body weight to steer the skis in the direction they wished to go. Though my grandfather didn't realize the full implications of it at the time, he was witnessing the earliest beginnings of the ski community in Utah.

During the 1930s and '40s, five ski resorts opened in Utah: Brighton in 1936, Alta and Snowbasin in 1939, Sundance, then known as Timp Haven, in 1943, and the predecessor to Deer Valley, Snow Park Ski Area, in 1946. Junior became involved in the development of some of these resorts, helping to build the second rope tow and the first lift at Timp Haven. He was a quick learner, and an even better teacher. Junior taught Maxine how to ski during their earliest days of courtship at Timp Haven.

He first caught the attention of the developing ski community in Utah when he competed in a cross-country race at Brighton.

No one knew who the teenager was, but after he beat out all the other more experienced competitors for gold, he inherited the nickname "plum picker" because of his farming background. Soon afterward he met the Engen brothers, Alf and Sverre, who had come south to help jump-start Utah's ski industry after helping develop Sun Valley. The brothers were drawn to Utah by the promising terrain and rumors that the snow there might be the lightest in the West. They weren't disappointed.

The Engen brothers took Junior under their wings. My grandfather was a fast racer and one of the best jumpers up Little and Big Cottonwood Canyons, but it was teaching that he really loved. Alf and Sverre trained Junior in the art of ski instruction, and in 1948 Junior became one of the first Americans to pass the Forest Service-sponsored certification exam in the Intermountain Ski Instructors Association. He and Maxine became instructors at Alta, living in the Rustler and Alta Lodges first, then building a home near Grizzly Gulch. They lived up the canyon for a few years, working as ski instructors during the winter and caring for the Provo orchard in the summer, before Junior was recruited to start the ski school at Sugar Bowl Resort in California.

When Ted Johnson began gathering the original investors to develop Snowbird, he insisted Junior return to Utah to build a ski school for the new resort and help design the runs. Junior stayed at Snowbird for the rest of his life, working there as ski school director and director of skiing until he retired at ninety.

Over the course of his almost eighty-year career, Junior became vice president and president of the Intermountain Ski Instructors Association, now a division of Professional Ski Instructors of America (PSIA), then served on the board of directors to create PSIA on a national level. He tested Rosemount boots, and designed, tested, and promoted Head skis, helping to create one of the first powder-specific skis. They were softer than the stiff mainstream skis of the time, and my grandpa and my father broke multiple pairs by taking them on one too many jumps. And in 1996, Junior was admitted into the National Ski Hall of Fame for his development of the American ski industry, his contribution to the modern-day powder-skiing technique, his dedication to teaching others, and his love for the mountains and snow.

When some recognize my last name, they look at me with a sort of reverence I don't deserve. *Are you related to Junior Bounous?* Yes I am, but no I'm not a ski instructor. Or a ski jumper. In fact, I'm so scared of jumping over cliffs, even small ones, that I'll hike back out of a situation I'm not comfortable with just to avoid them. And no, I don't race like my father. I don't coach either. I begin to see the slight confusion and disappointment in their eyes at this stage of the conversation. The last name Bounous carries a legacy within the mountain community here in the Wasatch. It's a legacy I've grown up in the shadow of, one that I'm proud to be a part of. I wonder sometimes what my role within this legacy is.

Settling

It was Christmas day. A storm had started the night before and continued until morning. Once we had finished opening our gifts, my cousins, my sister, and I pulled on our snow clothes, fitting our little feet into little boots and grabbing a few brightly colored plastic sleds from the storage room. The wafts of steam rising from the surface of the hot tub didn't tempt us that afternoon; our destination was the small hill just beyond the fence that surrounded the pool area. We postholed through the snow field, my older cousins helped us step over the fence, and we arrived at our destination: an untracked hill of snow, steep and short, bottoming out in the grove of pines below—perfect for sledding.

In the 1960s, Utah began using a marketing slogan to label Wasatch snow as "the greatest snow on earth." The implication of the slogan is that Utah snow is some of the "lightest" in the world, but according to atmospheric scientist Jim Steenburgh, author of *Secrets of the Greatest Snow on Earth*, Utah doesn't even have the lightest snow in the American West.

Snow's "lightness" is determined by water content. It's easy to look at a snow-covered hill, like the one we were sledding on that Christmas day, and assume that it's all snowflakes tightly packed together, but a fluffy layer of snow is mostly air. When snowflakes settle on a landscape, they are so light that they don't stack together tightly, and their six points and varying shapes create miniscule air pockets between the flakes. These shapes change and deteriorate over time, becoming more compressed as temperatures warm or when pressure is applied, like from footprints, skis, or sleds. After a snowfall in the Wasatch, the top layer might be light while the layers under have been compressed.

There are two main ways of measuring snow's lightness: the snow-to-liquid ratio and water content. The average snow-to-liquid ratio in the United States is around thirteen to one, meaning that thirteen inches of freshly fallen snow leads to one inch of water, though in Utah this ratio can be greater due to lower water content. Water content is measured by the percentage of water in the snow—what the depth of the water would be if a snowpack were to melt all at once. Light snow has a water content of 7 percent or less, average water content is 7–11 percent, and heavy is anything over 11 percent. The average water content of snow at Alta during a season is 8.4 percent, while snow at some resorts in Colorado is lighter, closer to 7 percent. The water content of individual snowstorms varies depending on certain conditions, including the temperature of an area before and during a storm, how much moisture is in the air, and what the wind is doing. Some snowfall in the Wasatch can be as light as 4 percent, while some can be over 15 percent, a percentage more common in snow found along the West Coast. Man-made snow, blown out of snow guns, is much higher even than this, usually 24–28 percent, which is good for filling in the landscape and creating groomed runs but doesn't lead to the fluffy conditions associated with powder.

Water content shapes the snowpack. A snowpack increases in water content as the winter continues, as older snow compresses and new storms drop fresh inches. The water content in a snowpack determines how much snowmelt there will be once winter passes into spring. By the time spring rolls around, a snowpack is likely to have a 40–50 percent water content. The higher the water content of a snowpack the more spring runoff will occur during the melting season, and the longer the water will flow during the summer months.

My ten-year-old self didn't know the water content of the snow we traipsed through, but it was so light it felt like clouds had fallen through the sky and settled between the trees that morning. Trying to climb back up the hill was nearly impossible. My legs disappeared in the snow with each step, so I'd throw my arms forward to try to spread my weight out, increasing my surface area to stay afloat. The going was slow but with each footprint the snow beneath would compress, creating a staircase to the top of the hill. And with each sled, the track became smoother and faster.

We sledded late into the afternoon, our pink noses runny, our eyelashes wet with sweat and snow. At one point, my eldest cousin Shanyn pushed my sister down a track so she could get more speed. Tyndall carried so much momentum she bounded past the end of the track, blasting through a snowbank that had formed between two trees and emerging with a face full of snow. Being the older sister, I wanted to up the game. Lying on the sled headfirst, I asked Shanyn to push me. I picked up speed quicker than Tyndall did, becoming more unstable as I approached the trees. Then I learned the hard way how soft snow compacts to become smooth and slick.

I got lucky. Instead of smashing headfirst into a tree, I collided with a patch of ice that had formed on the side of a trunk. My face shattered the ice and scraped against the bark it had covered. I was black and blue and bloody at Christmas dinner that night, my eye nearly swollen shut, my jaw aching with every bite, angry red scratches along my cheeks and down my neck. To this day, I have a dimple on my right cheek in a strange spot, higher on my cheekbone—not where most dimples form—a reminder of the painful differences between soft snow and hard snow.

Nesting

My grandfather Grant started our family tradition of staying at the Lodge at Snowbird for the holidays, a tradition that's been carried on for over forty-five years. Each holiday season we decorate an enormous tree, dangling over the upstairs balcony to precariously place bobbles on the topmost branches. In the evenings my mother and aunt play guitar while we watch the snowflakes drift outside the two-story picture windows, the gently flickering fire sending our shadows onto the walls. A motion-activated singing wreath hangs on the door to our rooms, facing the hallway, so that whenever someone passes the wreath yells, "Deck the halls with boughs of holly!" its fa-la-la-la-las echoing down the cold corridor long after the passerby has gone.

It was in these rooms that my parents officially met for the first time. Though the Bounous and Culley families knew each other from skiing at Alta, Snowbird's development brought the families closer. One Christmas, the Culleys invited the Bounouses over for dinner. My mother had returned from a season of working as a pastry chef at a heli-ski lodge in Canada, and my father had been named to the US Ski Team's development team. As my mother tells it, while she and her sister were doing the dishes, my father was sleeping on the couch, drooling. As my father tells it, he had been training every day starting at five a.m. for the past week, and was exhausted.

My mother's family hired my father to be their guide, and he took them to a chute called "Great Scott." Being a professional skier, he misinterpreted when they called themselves "advanced" skiers. The entrance into Great Scott is steep, and in early season conditions skiers need to step over exposed rocks to access it. Skipping over that step, my father jumped over the rocks, dropping into the chute and disappearing into the fog, leaving my mother and her family to navigate their way through the rocks.

On their last run together, my parents rode up a two-person chair called "Wilbere." As they approached the top, my father handed my mom a business card, telling her to call him up if she ever needed anything. The card didn't have his professional information on it, but instead listed inappropriate things he could be "hired" for. As they got off the lift, my mother said, "You're overpriced."

To this day, she still keeps that card in her purse.

The Gaylords don't own the Cliff Lodge anymore, but they still gather in Salt Lake City during the winters. Colin's parents live on Orcas Island, part of an archipelago called the San Juan Islands off the coast of Washington State, and fly to Utah every winter for Christmas. The island is where Colin and his sister grew up. Coincidentally, my mother grew up attending a summer camp on Orcas Island, working there when she was older. My sister and I attended the same camp and eventually became counselors and activity-area heads. Years before I met Colin, I worked as a counselor there with his younger sister, Genevieve. And during the one year I took off, between when I was a camper and a counselor, Colin worked as a trip leader at the camp, taking campers and counselors on sailing trips around the rocky shorelines of the San Juans. Colin's mother's side of the family still lives on Orcas. The island is another landscape that we share.

My family invited Colin's immediate family up to our rooms at Snowbird one night for dinner over the holidays. I brought up what Bud had told me the prior summer, that the Gaylords used to own the set of rooms next door. Colin's mother, Marny, laughed.

"You know, I remember those rooms! That's where Randy and I first met."

Marny had been invited to a Gaylord family Christmas by her college roommate, one of Randy's sisters. It was the notorious season of 1976/77, considered the worst snow year on record until the 2017/18 season. The ski resorts didn't have the ability to make snow yet, and weren't able to open the lifts. So the Gaylords hiked rather than skied, and it was during those hikes that Randy and Marny got to know each other. Since Randy grew up racing and was a stronger skier, they admit that they may not have gotten to know each other if the resort had been open. There is, of course, the possibility that even if the snow had arrived in time for the holidays, their paths crossing in those rooms may have been enough to ignite the spark between them, but there is also the chance that their relationship may have never started taking shape.

During their story I reached over and placed a hand on Colin's leg, giving it a little squeeze. It was here, one year ago, that Colin and I started getting to know each other, too. Having been introduced the prior fall by a mutual friend, it wasn't until the snow started falling that we really got to know each other. On Christmas that year, Colin and Randy met my dad and me for a few runs. It was a powder day, and as we rode the lift together our fathers discussed their days of racing on the Snowbird Ski Team together. Colin and I sat next to each other, occasionally asking each other questions in oddly polite voices.

During our first run my father correctly predicted that ski patrol would drop a gate, and we were treated to fresh tracks on a run called "Eddie Moe's." We took our first powder run together, snow billowing off our chests. We reached the bottom, sharing laughter at such an amazing Christmas day, both pleasantly surprised with how well the other skied. With each run our conversations became more relaxed, our voices more prone to laughter.

Stacking

There were mornings when the first thing I became aware of was cold air falling from the window just above my bed, the chilled air brushing the exposed skin on my face. My heart would jump a beat in excitement. I'd pull the curtains open to see a landscape of white, snow stacked precariously upon the bare branches of scrub oak. Where there had barely been an inch, or less, on the porch the night before, there would be a solid two feet of snow.

If it was a weekend and my dad didn't have work, my sister and I would crawl into our parents' bed, disrupting their peaceful moments together. They had the best view from their lofted room; their two-story picture windows looked out at our tiered yard and the mountains behind it. But we didn't have eyes for the view, just for the implication of what snow meant: bobsledding.

Since our home is built into a hillside we have a three-tiered lawn, which means a three-tiered bobsled run with banking turns expertly designed and executed by a man who understands momentum and snow. While our father did the manual work, Tyndall and I would entertain ourselves by throwing snowballs at each other or at him (he'd retaliate by dumping a shovelful of snow on our heads), and then spend hours hiking up to sled back down. Afterward we'd be rewarded with a fire, stovetop popcorn and hot cocoa, courtesy of our mother.

If it was a weekday, my parents would already be downstairs, my mother portioning scrambled eggs onto our plates. Our question was always the same: Snow day?

And the answer was always the same: No.

Every so often, usually after at least a foot if not two feet of snowfall, the local public school system might close for the day. But Tyndall and I attended a private school. Ours did not have to follow Granite School District decisions, and as a result we never reaped the benefits of heavy snowfall; in my twelve years attending the school I never had a single snow day.

Because we also grew up in a private neighborhood, which the city didn't plow, when we got two feet of snow it meant that our parents, and us when we got old enough, had to drive through our unplowed neighborhood to get to the nearest plowed road. Our neighborhood was very hilly, the streets lined with scrub oak. A couple times each winter, a car would slide off the road and get stuck enough that a tow truck had to pull it out. One of the steepest hills in the neighborhood was our own driveway. As a child I got used to the way our cars would slide out at the bottom of the hill, coasting toward the oaks before my parents could correct them.

Since I turned sixteen in January, my first few months of driving were fraught with sketchy experiences while driving in the snow. The mildest of these were when my all-wheel-drive car would slide just a few inches off the road at the end of our driveway, or slide backward when I didn't have enough momentum going up it. Sometimes my best friend, Whitney, wouldn't even be able to make it up my driveway when she'd visit. The more alarming instances included another hill in our neighborhood, one that dead-ended with a stop sign and T-boned another street. If I started sliding at the top of that pitch, all I could do was hope that someone wasn't driving on the other street at the time.

Since I was one of the first of my friends to get my license, and a member of our school's volleyball team, I often ended up driving after team activities. Once, trying to get half of the younger members of the team home before their curfews, we squeezed thirteen girls into my Toyota Highlander. Right as we got on the freeway a blizzard began, erasing the lines on the dark road with inches of snow. I couldn't tell where the shoulder was, or if we were even in a lane. The snow was falling in such thick, fast chunks that even if the lines had been visible it would have been nearly impossible to see them.

It was a valuable lesson in more ways than one. I discovered that turning on the car's brightest lights decreases visibility, making it seem as though the vehicle has taken off into warp speed. I also learned, quite the hard way, that the way to defrost a car window is *not* by blowing cold air on the windshield. When the windshield wouldn't defrost after Whitney, riding shotgun, had turned the cold air on high for a few minutes, we decided it would be best to open all the windows of the car to try to get them to clear up. I wish I were joking. We spent fifteen minutes with the car windows rolled down in the middle of a blizzard on the freeway in the dark with thirteen girls out of seatbelts stacked on top of each other. Perhaps it was lucky that the storm moved in; it forced me to drive slower than twenty miles per hour. Some of the girls did miss their curfews, but they all made it safely home. And Whitney and I eventually figured out that warm air, not cold, is how you defrost a window.

In places that get hit with heavy winters, dealing with snow can be a real challenge. Since Utah experiences consistent snowfall through the winter months, even during low-snow years, our metropolitan areas are well-equipped to deal with snowfall in the valleys. A few inches of snow rarely halts daily life, like it might do in less prepared places like Seattle or San Francisco. When a storm drops closer to a foot of snow, movement in the Salt Lake Valley can become more difficult. I've left my house at seven a.m. to go skiing when the snow on the streets is as high as my tire wells and there are few other cars in sight, even though it's a weekday and should be prime traffic time. It's only as I approach the entrance to the canyons that the line of cars begins to stack up.

The canyons are a different story. Steep pitches and scooping valleys create avalanche hazards when the snow starts falling. The canyons will close for avalanche control work, often at night or early in the morning, sometimes in the midafternoon if the avalanche danger is high. Usually ski patrol and UDOT workers can get the canyon open after some control work, but sometimes the canyon can close for hours or more. Occasionally patrons and employees have to spend the night at whichever resort can take them. In 2019, an avalanche slid across the road after the canyon had reopened after control work, while cars were driving down. When we told one of Colin's uncles—who used to work as ski patrol at Snowbird—about the avalanche, he responded by saying: "I love it when nature flexes." Our climate and weather patterns define how we live in a landscape. I, too, like being reminded of that.

Living in the snow is not easy. It requires effort. Before snowplows, citizens would take to the streets with shovels to level the snow on the roads for horse-drawn sleighs. The earliest snowplows began appearing in the Midwest and Northeast in the mid-1800s. They were pulled by horses and allowed society to continue to function during the winter, even during periods of heavy snowfall. But snowplows had their drawbacks. Snow doesn't just disappear when it gets pushed off of a road; it has to go somewhere. Snow piled up along the sides of roads, posing issues for pedestrians. When snowplows came on the scene they were initially praised, but public opinion flipped when large berms of snow blocked local businesses or cut off pedestrian access. People on foot had to hike up over these mounds just to cross the street. And because the piles were so large, they didn't melt even when the weather warmed up and the sun came out. Methods of snow *removal* had to be explored, so snow would actually be removed from an area entirely rather than pushed to the side. City officials often hired teams to take snow out of cities and dump it into rivers or in rural areas.

This is an enormous undertaking for cities to this day. I was staying up in Park City, a smaller town on the backside of the Wasatch, during one thrilling storm cycle. The snowfall was so consistent and so dense that there was little reprieve from the blizzard for days. The roads there are small and narrow, and some are quite hilly. Driving through the city became hazardous, and when plows moved the snow around it would stack up on the sides of the roads, making the streets even narrower. The city began bringing in dump trucks that would relocate the snow to an empty plot of land outside of the city, where a plow would move the snow to the far side of the plot, building a large, wedge-shaped pile.

We can get intense storm cycles here in Utah, but they often don't compare to the amount of snow that can fall in the Sierra Nevada, where more humidity in the air leads to higher amounts of precipitation, or to the icy intensity of the storms that can slam the Midwest and Northeast. While the snowfall here is, for the most part, light and fluffy and melts quickly when the sun reemerges, the storms that freeze other regions of the country can be nasty and even deadly.

When these regions are hit with intense winter weather, many people scoff at climate scientists, claiming that these storms are proof that the climate isn't warming. But the brutal ice storms that originate in the Arctic and blast the Midwest and Northeast, commonly called the *polar vortex*, are actually side effects of carbon emissions altering the *jet stream*.

The jet stream is essentially a river of wind, about five to ten miles above the earth's surface, that travels across the upper and lower quadrants of the earth, encircling each pole. The northern jet stream moves from west to east at speeds up to 275 miles per hour, wavering north and south slightly as it circles. When the jet stream strength weakens, the cold, icy air—the polar vortex—

being held above the pole "leaks" down into lower latitudes, bringing Arctic weather down into eastern North America. This also allows warmer and potentially drier air from around the equator to "flood" north in latitude, affecting the western part of the continent and contributing to warmer, drier winters.

Climate change is causing the jet stream to weaken in a few different ways. Arctic ice melt is causing the oceans to warm up, which in turn leads to more melting ice. This affects the temperature gradient around the poles; the jet stream is driven by the radical temperature differences between poles and equator. And warming oceans in the tropics and the El Niño weather pattern that occurs during certain years are other factors that could be weakening the atmospheric river's path.

When winter weather patterns occur, some climate skeptics take advantage of the cold conditions to claim that climate change isn't happening, while basically admitting that they don't know the difference between climate and weather. While this is frustrating, it doesn't concern me too much. What I worry about most is when people who I consider "climate neutralists"—those who believe in climate change if you ask them, but who are neutral when it comes to changing political policy and taking action against oil companies—seem to forget that climate change exists during heavy winters. I've noticed that when a snow season in Utah is bad, members of the ski community will have more conversations about climate change. But when a winter is good, with consistent storms bringing inches and feet of snow to the mountains and valley every few days, those conversations go silent. I don't know how to carry the climate momentum that can develop during a terrible winter across seasons and years.

Swirling

The smell in the tram was atrocious. A mix of wet sweat and body odor, it was almost as claustrophobic as the people themselves, crammed together like sardines in a tin, rubbing shoulders and skis as we swayed back and forth. Colin and I were usually pretty good at either being first or last in so we can stand next to a window, but a last-minute group squeezed in behind us, forcing us further from the window and deeper into the pit of body odor.

There was a final push as the tram attendant slid the door shut, and then we were moving away from the deck, into the open air above the slopes.

"It smells disgusting in here," I mumbled to Colin.

He raised his eyebrows and shrugged. "Powder day!"

My own inside layers were soaked through from the efforts of the day, and I knew I couldn't smell much better than everyone else. A conversational hum started as the mountain changed shape beneath us, becoming steeper and speckled with pines and cliff bands.

"Are there any chicks in here?" someone asked with a laugh in his tone, his voice projecting through the tram. I glanced in the direction of the voice, caught the playful smile on his face as he scanned the crowd. Our eyes met and he turned away, smile fading a little, embarrassed. I could feel eyes on me as goggled heads turned to see if they could answer his question. A few moments passed, awkwardly, before the typical excited buzz of the tram resumed. Catching Colin's grin, I gave a small shrug. It wasn't the first time I had been one of the only women on the tram during a powder day.

The clouds thickened as we climbed in elevation. The trees and runs below dissolved into the all-consuming white. The lumbering piece of metal swung as it passed each tower, and the bodies inside swung with it. After what felt like half an hour, though it was only a few minutes, the tram slowed, a sign we'd finally come over the crest of the mountain and were approaching the deck. A fierce wind started that we could feel even through the heavy weight of the tram.

"I think we should go hike Baldy," Colin said quietly in my ear as the murmurs in the tram died down again. "I bet the wind's made it good."

Becoming intimate with a mountain's terrain can be useful on days like today. Microclimates and variable conditions occur all over a mountain. During a whiteout, with the wind coming from the southwest, it would be suicide to drop into a south-facing area like Mineral Basin or Powder Paradise. A decision to venture there could mean a few good turns, but it could also mean twenty minutes of absorbing cut-up crud you can't see. Exposed ridgelines are misery for your face, and potential frostbite. Runs with no trees cause vertigo. If the wind's coming from the northwest, it can sometimes mean unexpected inches from lake effect.

The direction the wind was coming from today meant that some areas were scoured, the fresh snow buffed out and shredded to the icy layers beneath, while other features were actively filling with the flakes that had been blown off ridges. Our last few runs were on Great Scott, a north-facing chute that refills with windblown snow all day—the same that my young father abandoned my mother's family on. The snow was firmer and chalky at the top and deep and soft toward the bottom, conditions that Colin and I love, but between exposed rocks at the entrance, low visibility, and crowds, we were ready for new terrain.

Baldy, the peak just to the east of Hidden Peak, has runs that benefit from the same windblown phenomena. Instead of stepping over rocks, however, you have to click out of your skis and hike along a ridgeline. I had expected Colin to suggest Baldy next. It's our go-to in conditions like this.

The tram doors slid open with a chilling gust of snow and wind. I kept my head down as I blindly followed the pairs of boots in front of me, feeling more like a sheep during these moments than almost any other time in my life. Within a few minutes we had stepped into our skis, coasting along a cat track that would take us to the gates of Baldy.

The hike is along a ridgeline that, though not terribly exposed, passes close to the openings of some chutes on the right. I once was hiking behind a guy who stumbled near one of these openings and almost lost his skis down a chute. When combined with wind, deep snow that can send a boot off balance, and a whiteout, passing these chutes can become precarious.

Using ski poles to steady myself against the wind, I started off briskly, following close behind Colin as though his body could act as a windbreak. Each step was a reach for me. I stumbled a few times, cursing whichever long-legged ski patrolman broke this trail. My nose burned from the frigid wind, but each time I tucked it under the buff around my neck the moisture from my breath fogged my goggles up. I spent the hike switching between risking frostbite on my nose and hiking blindly with fogged goggles. A series of steeper steps caused me to fall further behind Colin, and I quickened my pace to catch up. I was panting so hard from the effort I was almost hyperventilating. I began timing each breath with a step to regain control.

Inhale . . . exhale . . . inhale . . .

The trail left the ridgeline, cutting north beneath the peak's summit. We reached a bench-like feature on the slope where others were placing their skis and boards on the ground, knocking snow off their boots and strapping into their gear quickly and quietly, a sense of urgency in their jerky movements. There was a sort of mania to the scene, and Colin and I wove through the small crowd, trying to find space to buckle in. This throng of enthusiasts would disperse as people skied down the traverse, when the mountain opens up and skiers and boarders can choose their own course to the bottom. But the cramped space where hiking trail meets traverse can force more skiers together on a powder day than almost anywhere else on the mountain, besides in a lift line or the tram. The payoff for this hike is three thousand vertical feet of some of the steepest terrain and deepest powder on the mountain.

Hiking Baldy is not always this manic. On sunny days I'd go so far as to say that the hike can be pleasant, even *easy* if the snow has been packed down enough. On those days I can walk with my own stride, not having to follow in the footsteps of others. Sometimes we'll hike Baldy just to appreciate the view, no matter what the snow conditions are. If it's bad snow, breakable windcrust or ice, it's a long, miserable run to the bottom, but we risk it to have a few moments on the bench.

Once, on a powder day, we stopped our hike when we saw a porcupine in a tree. Powder be damned, we abandoned our skis on the side of the trail and postholed closer to the tree in feet of untouched snow. We must have spent thirty minutes watching the porcupine gnaw on a branch, while hordes of people continued up to get the fresh tracks we were forsaking. Colin, who had never seen a porcupine in the wild, was enthralled. We joked that he had finally met his spirit animal: equally prickly and lovable.

On clear days, Baldy is the best place to witness the passage of the sun across the landscape. Directly across the canyon from Baldy, Mount Superior's long, south-facing ridge lengthens as the sun passes its apex, the resulting shadow snakelike as it curves from the tip of the peak to the canyon road. The silhouettes of the highest ridgelines and peaks become sharp and jagged against cerulean skies. The shadows of pine trees lengthen and turn blue on white-gold snow, while crags deepen and darken in cliff bands across the canyon. Little Cottonwood Canyon's depth is revealed as particles in the air scatter the receding light of the sun, making the canal that connects mountains to valley floor misty. The Oquirrhs become a distant, faded blue across the valley.

Occasionally one of us will pull a beer out of a pocket. I'll lean my helmeted head awkwardly against Colin's shoulder, craving his touch even through ski clothes. My dream is to one day bring a backpack with a thick wool blanket and champagne, and take an hour to just sit there and watch as the winter sun sets over the golden ridgelines of the Wasatch. When friends visit we always make an effort to bring them to this place. Since most of them live at sea level they might fall a little behind on the hike, struggling to bring in enough oxygen in the thin air, but they never complain. They reach the bench and have to take a few moments to absorb everything. As they look around at their surroundings, I look at them. I like watching their smiles as they arrive at my favorite place in the Wasatch.

Sometimes, for a very brief moment during that quiet reverence that falls as we watch the peaks glow, I feel like I'm getting a glimpse of something, just out of reach. I don't know what it is, but before I can hold onto the moment we set off down the hill and it flies away with snow from our skis. All I take with me is that this snow, the mountain—or is it the man?—is intoxicating.

Once, while passing a Mormon church in the valley on a Sunday morning, the parking lot overflowing with cars, Colin said, "What do people see when they go to church? I don't understand how anyone under fluorescent lights can say they've had a spiritual experience."

"What would you consider a spiritual experience?" I asked.

"I don't know."

He was silent for a moment, watching a family dressed in stiff Sunday clothes walk through the church doors.

"I don't know," he repeated, "but I'd say hiking Baldy is pretty spiritual for me."

They say that timing is everything in a relationship. But I'm beginning to think that place has a lot to do with it as well.

Fracturing

The snow that my grandfather, Grant, loved so much was once an inch away from killing him when he was caught in an avalanche.

My grandfather, two of his friends, and my mother, who was only twelve years old at the time, had taken a helicopter from Alta into a neighboring basin. It was 1968. The slope that the avalanche occurred on would be developed into a run called "Regulator Johnson" when Snowbird opened up a few years later, but at the time it was backcountry terrain.

Perched halfway down the pitch, they decided that my grandfather would be the one to get first tracks. He cut sideways across the hill, took one turn, and then he and the mountainside disappeared in a blast of wind and a churning wave of white.

My mother had been standing uphill from him when the snow broke and the hill cracked open. The force of the air that escaped the spaces between the snow crystals as the snow compressed on itself was so strong that my mom was blown over on her side. By the time she could look around, the avalanche was coming to a stop some two hundred yards downhill. My mother stood on a precipice; though one ski was on the stable snow beneath her, the other dangled some three to four feet above the freshly exposed layer. As the snow had fractured beneath her, her family fractured before her eyes. Her father was gone.

My grandfather's friends lowered themselves onto the layer of snow left behind, leaving my mother perched on the newly made ledge. Wild with fear, she managed to get down onto the slick avalanche path. The avalanche had gathered in the basin, the powder snow now bunched into balls of ice. Once the air pockets between snow crystals is forced out, snow compresses, becoming heavy and dense. As an avalanche loses its momentum and slows, its consistency becomes cement-like.

A common reason that avalanches occur is from a "weak" or "rotten" layer. With every storm, warm front, clear and cold night, and significant wind current in the mountains, the topmost layer of a snowpack can change. When that altered layer gets buried by new snow, it can become an avalanche threat.

A common type of snow that creates a weak layer is *surface hoar*. Surface hoar, essentially frost, is one of the most beautiful types of snow formations. Feathery in appearance, it develops on the top of the snowpack and looks like frozen starlight. On clear and cold nights, snow radiates heat away from itself, becoming colder. Cold air sinks to the ground at the same time, becoming more humid as it does so. When the cold moisture meets the snow, moisture latches onto crystals and freezes to create those feathers, which often stand straight up towards the stars.

Surface hoar is hard to track. Sometimes it might occur in open spaces, avoiding tree groves. Sometimes it might appear near the bottom of a mountain, where cold air is likely to settle. Other times, it might be found at the top of a mountain if there was lower-elevation fog. If any wind travels above hoar the feathery crystals take flight or collapse, creating unpredictable pockets where some hoar has disappeared and some remains.

Thin, slippery, and fragile, surface hoar often doesn't remain on the surface for too long before new snowfall covers it. Compared to potato chips by some avalanche experts, hoar can take a certain amount of weight before it collapses under pressure. When surface hoar shatters, the air trapped in the layer is forced out, causing the snow to drop like a jerking elevator. The compressed layer then becomes an icy slide for the snow on top, gathering more snow and momentum as it travels downhill. The fastest avalanche ever recorded was on the side of Mount St. Helens when the volcano erupted. It clocked in at 250 miles per hour.

There was no hint of pole or ski or hat or Grant Culley. Just by chance, one of his friends saw a puff of snow blown into the air, diamond dust catching the sun for a moment before settling back down into the wreckage. My grandfather had been frozen in an upright position, head bent back with only his nose poking out of the snow. Once the avalanche stopped moving and cemented him in place, he was able to tell from the lightness of the snow by his face that he was near the surface. So he did the only thing available to him: blow through his nose as hard as he could. After a few bursts he was able to breathe fresh air, and the resulting puffs of snow from his nostrils alerted his friends and daughter to his location.

It took hours to dig him out. The group hadn't been carrying shovels, so they used their skis to slowly break apart the solid snow that held Grant in place. Besides one of his legs, which had been cut up by a tree he collided with during the slide, he was miraculously okay. If he had been buried any deeper, he likely would have died, since no one had avalanche beacons that day. He tried to keep the avalanche a secret from my grandmother, but when he limped into the hotel room she could tell something had happened. To this day, my mother can't talk about the event without choking up.

Snow has helped to create and sustain most of the relationships in my family, yet the mountains pose threats to those who enjoy them. Avalanches, though rare, take the lives of Utahns almost every winter, typically those who venture into the backcountry either on skis, snowboards, snowshoes, or snowmobiles.

My parents and paternal grandparents have all been caught in avalanches that they've been able to ski out of. Once, prior to active avalanche control work in the canyon, Junior and Maxine were driving down Little Cottonwood Canyon when they came across an avalanche that had crossed the road They turned around, thinking to stay the night at one of the lodges at Alta or Snowbird. But before they could reach the resorts they came across another avalanche blocking their path, which had released in the time since they had driven past. One year when they were living at Alta, a series of massive avalanches closed the canyon road for two weeks, and food and supplies needed to be air-dropped to the community.

In 2015 a young man from my dad's ski team was lost in an avalanche while training with the US Ski Team in the Austrian Alps. His name was Bryce Astle, and he was two weeks shy of twenty when he died. Bryce and his two brothers grew up skiing at Alta. They spent their winters as the local ragtag crew of skiers, turning from boys ripping around groomers to teenagers sending backflips off of cliffs. But no one was quite like Bryce. He became such a talented skier that, on the cusp of adulthood, he caught the attention of the US Ski Team.

We went over to his family's house on the morning of his death. Before we had a chance to walk inside, we ran into Bryce's older brother in the driveway; he was loading his and his younger brother's skis in the car. His eyes had an expression that I'll never forget. Standing in front of his home, he looked completely lost. After giving him our condolences, we asked if he was heading up to ski. He looked up to the mountains. He nodded, no longer looking quite so lost.

"That's the only thing we can do right now."

Racing

"All set? All right, let's head out."

Unsuccessfully trying to stifle a yawn, I push my skis into motion, gliding down behind the other members of course crew. My father leads the charge, skating expertly up to the lift with an enormous bundle of gates on his shoulder. He stops at the lift, looking behind him. Realizing he's waiting for me, I skate up to his side, feeling quite weak and pathetic carrying just my small squirt bottle filled with blue powdered dye. He gives me a semi-patient smile as the lift swings behind us, hoisting us into the brisk dawn.

The mountains are just beginning to light up with alpenglow, stripes of amethyst and rose as the first sun rays bend around the curve of the earth to hit them. Turning my head, I try to see the first golden light touch their peaks, but the predawn breeze makes my eyes water and feel as though someone has placed a block of ice against my forehead. Following my father's lead, I lower my ski goggles reluctantly.

As director of the Snowbird Ski Team, my father is "Master of Race" for most ski races that take place at Snowbird. If he needs extra hands to set up gates or nets, or, in my case, to sprinkle dye at the base of each gate, I'll occasionally drag myself out of bed to get onto a ski lift before the sun has risen. Sometimes it feels as though I might be more hindrance than I am help, since I have no background racing or setting up gates, but I enjoy the privilege of putting on my skis before the rest of the public does, even if my runs consist of slide-slipping through the race course. Sitting on the lift with him in the predawn sky, watching the mountains turn from deep indigo to a golden scarlet, is a kind of peace I can look back on when life gets too loud and fast.

Despite his background in racing, my father never pressured my sister and me to race, though we were required to know how to ski, despite our best efforts. My parents put me in the youngest race group for one year before I demanded out. I say one year, but it was probably more like one week. One of my father's coaches tried to correct my technique, and, being the stubborn, snobbish six-year-old I was, I responded, "You can't tell me what to do, my dad is your boss!"

I think the coaches were just as excited to have me out of the program as I was.

As I got older, the thought of racing crossed my mind quite a few times. But I became afraid and embarrassed even at the thought of trying to run gates in front of my peers and coaches, knowing that everyone would expect me to be a good racer and freeskier, like my dad and grandfather. So, to avoid failing other people's expectations—to avoiding failing *my* expectations—I just didn't race.

There are moments when I regret it. Like when I sit on a lift with my father in the dark before the dawn, and think about all the sunrises I could have witnessed with him during early morning training.

We push off the chair and coast down a cat track. He picks up speed much quicker than I do, between the sharp edges of his race skis and the added weight of the gates. Not as accustomed to skiing without poles as he is, I angle the inside edges of my skis against the snow and push them away from me one after the other, gaining forward momentum as though on a pair of skates, attempting to catch him before he gets too far away. Even in my late twenties I get anxious when I fall too far behind him.

Aware of coaches and racers on the chairlift above me, and that my father and I are the first ones on the hill, I'm extra attentive to how much pressure I apply to the front of my boots, how forward my hips are, how the edges of my skis carve into the untouched corduroy. Looking like a top-heavy triangle of sorts, my dad makes beautiful, arcing turns across the entire hillside, as stable with fifty pounds of gates on his shoulder as with no weight.

He picks up momentum as the hill steepens, utilizing the extra weight to increase power in his turns, harnessing the energy of the pitch and his skis and the snow. I try to follow, but as I pick up speed I become fearful. I maneuver my feet, partially cutting off my momentum. In the breath it takes me to brake, my father reaches the bottom of the hill, dashing down a cat track that leads to the top of the race hill. I follow as best I can, trying to still my anxiety at being left behind and find a balance between strength and speed.

Since I don't have any experience drilling gates, once the race begins my dad assigns me to watch gates. Officially, I am a gatekeeper. My duty is simple but critical: make sure the racers go through my assigned gates correctly, or mark them down to be disqualified.

A radio crackles to my left.

"We've got a gate leaning in toward the hill after the hairpin."

"Stop start. Course hold."

"Start is stopped. Bib 23 on course, Bib 24 is in the start."

My dad watches as Bib 23 *thwaps* past, then skates out onto the course. He drills another hole a few inches away from where the leaning gate was, pulls the plastic stick out of the ground, and shoves it down into the new hole. Pressing one of the four radios decorating his chest, he reports, "Start is clear to resume."

"Start resume. Bib 24 in the gate. Bib 24 on course."

And then from another radio: "Bea on."

A few seconds later, Bib 24, looking impossibly tiny but determined, comes peeling over the hump that hid the start of the race from us. Her legs look no thicker than my forearms in the tight pink-and-white speed suit, and her golden hair flaps behind her like a miniature cape.

My dad mutters under his breath as she cuts close to a gate, avoiding a rut that's thrown a few of the racers off balance.

"Arms up, Bea, get those arms up!"

We can hear her small, incoherent voice as she gets closer, talking to herself, perhaps saying the same thing that my dad is muttering. As she *thwaps* the gate in front of us, he raises his voice.

"Hup hup hup Bea! Hup hup hup!"

Bea catches a ski tip on one of the gates below us, flipping her around backward as she stops a feet few beneath the gate. The coaches and spectators groan, then start cheering.

My dad presses his chest.

"Bib 24 is hiking."

Flailing pink-and-white limbs, neon yellow poles, and sharp skis chop madly back up the hill. She pushes herself around the gate and back onto the track. Though her mistake costs her time during that run, tomorrow Bea will get third in her age group.

With envy I watch my dad cheer for her like he had never been able to cheer for me as a racer.

It would've taken me on a different path, I tell myself, resorting to the thought that comforts me when I consider the alternate life I could've had, one that would have meant chasing winter throughout the year. *I wouldn't be where I am today. And I love where I am today.*

Another tiny body *thwaps* past, and an assurance comes to mind.

He'll have grandkids who will want to race. He'll get his parent-coaching fill then.

I try to imagine what my kids will look like, their limbs in tiny pink speed suits, my dad muttering under his breath while watching his granddaughters rip around gates.

That's when the tears start coming.

I will never be a mother.

Sinking

I remember the January of my senior year of high school vividly, not because of any certain events, but because of the air quality. I remember that January because I couldn't see the mountains at all, despite my high school campus having mesmerizing views of the Wasatch. The air was hazy and pink, sometimes purple, orange, or grayish-brown. It worsened with each passing day, the pollution compressing and thickening the longer the valley went without a storm. The sky, tinged with murky colors, darkened as layers of pollution blocked the sun. The temperature was frigid, sometimes dipping down into single digits. My high school campus was an open campus, a cluster of one-story buildings spread out across a few acres, and we had to walk outside between classes. As the air quality worsened, the teachers stood by the doors to hurry students inside, opening and closing the doors as quickly as possible to try to preserve the quality of the indoor air. When we'd get home, we'd hear the local news stations warning not to spend any time outside if we could help it, that breathing the air was worse than secondhand smoke, that pregnant women, children, elderly folks, and people with asthma and heart conditions were especially at risk. Hospitals reported a thirty percent rise in visitations. My father would return home from working up at Snowbird and relay that up in the mountains it was almost fifty degrees and sunny, a perfect bluebird day.

This phenomenon, known as an *inversion*, gives us the nickname "Smog Lake City." When cold temperatures, carbon emissions, and a stagnant weather pattern occur, the Salt Lake Valley can swiftly become one of the most polluted cities in America, and even in the world. Though inversions occur naturally in valleys, when they combine with human-caused pollution they affect the quality of life in the Salt Lake Valley during the winter months.

High-pressure systems are the root causes of inversions, occurring when there is a higher concentration of atmospheric pressure above the earth, which forces air away from the center of the pressure system. This happens in a clockwise motion in the Northern Hemisphere, and pushes colder air to the ground while transporting warmer air aloft. In low-pressure systems the opposite occurs. When there is less pressure above the earth, it creates a sort of vacuum that pulls air up toward the atmosphere in a counterclockwise pattern. During a low-pressure system, moisture collects near the earth's surface, leading to clouds and sometimes rainfall as moisture condenses with pressure. High-pressure systems typically mean no clouds or moisture. The low angle of the sun during the winter months also means that there is less thermal energy to heat the air mass.

Inversions can occur after a low-pressure storm system has moved through the Wasatch, allowing for a high-pressure system to take place. The effects of an inversion are increased if there is a layer of snow on the ground, creating a cold, denser layer. During a normal weather pattern, warmer air rests near the earth while colder air remains higher in the atmosphere.

When an inversion occurs, this striation switches: the layer of warm air that would typically be closer to the earth is suspended in the atmosphere, sandwiched between two layers of cold air, one closer to the ground and the other higher in the atmosphere. When this happens in a valley or depression, the cold air can get trapped in the feature, the warmer air above it creating a "cap" that prevents the cold air from rising. This is why the mountains can be warm and beautiful while the valley is cold and miserable. The inversion worsens as time goes on, the denser, cold layer compressing and building as more pollution is released into the air, until another weather pattern moves through to mix the layers up. When an inversion occurs around an urban area, whatever is mixed in with the air particles—like harmful toxins and dust—gets trapped with the cold air as well, leading to discoloration and haze.

The intensity of the inversions during the winter is also influenced by the snow that season. Though snow in the valley can exacerbate the inversion, during big snow years, when there are storms constantly moving in and out of the valley, inversions don't get the chance to build up and become bad. When a healthy snowpack is able to develop up the canyon, the valley is able to maintain healthier air levels. More powder days mean more "green-air" days.

The mountains provide an escape from inversions, but only for those with access to the right transportation and freedom from work or school to leave the valley. From the peaks, the inversion simply looks like a thick layer of fog, not the stagnant underworld it becomes once immersed. The mountains during the winter create a "pull" that draws people from the city to the resorts, and, according to a study conducted by Utah State University, the inversion also creates a "push" that can cause an influx of visitors to the resorts during periods of pollution. The increased traffic to the mountains results in more carbon emissions and particulate matter, adding to the inversion.

Salt Lake City's inversions are caused by car emissions, commercial industries, and homes and buildings. Like most environmental problems, it's a complex issue that involves consumer behavior, politics, and industrial progress. It's also an issue of environmental justice. Many of the industrial sources of pollution are located where they impact low-income areas and communities of color more than wealthier neighborhoods. And there is an overall stagnation when it comes to creating change, a political gridlock that traps the pollution in the valley during the winter.

Some politicians will claim that the inversion has improved in recent years. This is true in some ways. My grandparents remember inversions worse than the ones the valley experiences today, with even higher levels of toxic particles. More stringent regulations on car emissions, building emissions, and industries have led to less pollution emitted per person, yet because the population of Northern Utah will continue to grow the inversion will continue to worsen, unless more regulations are put in place to counteract the growth.

The most common pollutant during inversion season is particulate matter, or PM. PM can be almost anything, from smoke to dust to drops of liquid to metals, and is measured in micrometers. PM_{10} are larger particles, measuring 10 micrometers across or less, able to be inhaled into the body. The most common PM_{10} include pollen, mold, and larger particles of dust. These the body can eliminate, trapping the larger particles in mucus before they can get too deep into the lungs. PM_{10} concentrations contribute to reduction in visibility, but are not as much a cause for concern regarding human health.

Smaller particles, measuring 2.5 micrometers or less, are the real threat during inversions and are divided into two categories: primary particles and secondary particles. Primary particles are released directly from the source of pollution. These often originate from wood fires, open pit mines, and fuel combustion from vehicles and industry, which release carbon monoxide, nitrogen dioxide, sulfur dioxide, and lead. Secondary particles are products of chemical reactions. When toxins, like nitrogen dioxide and sulfur dioxide, are released into the atmosphere and meet other toxins or molecules like water, they undergo chemical reactions, producing secondary particles. Nitrogen dioxide creates ozone, the main pollutant in the Salt Lake Valley during the summer months, and both nitrogen and sulphur dioxide produce acid rain when they react with water.

$PM_{2.5}$ can travel deep into the lungs, aggravating respiratory organs already sensitive to conditions like asthma and chronic obstructive pulmonary disease. The tiny particles also enter the bloodstream and travel all through the body. When they enter the heart, they can cause heart failure and coronary heart disease. If they enter the brain, they can impair cognitive function in adults and hinder a child's cognitive development. Some studies have found links between exposure to air pollution during pregnancy with autism and Alzheimer's. Even healthy individuals without preexisting conditions can be affected by $PM_{2.5}$, experiencing coughing, chest tightness and shortness of breath, and eye, nose, and throat irritation. While some of my friends struggle to breathe during inversions, I suffer from stinging eyes and headaches.

Each winter new studies are published about the health hazards of inversions. Pregnant women are advised to buy face masks and remain inside, and warned against taking babies and children outside. I have a friend who rented a condo in Park City during the winter so she could escape the inversion during her pregnancy. I would do the same, but many who live in the Salt Lake Valley don't have the financial means to rent a second home. In the past few years, it's become more common to see residents wearing face masks as they walk outside or use public transit. Even remaining inside can't keep us from the harmful effects of the inversion; doors and windows only do so much when it comes to $PM_{2.5}$. Last winter my parents bought air sensors, placing them around the house to measure the air quality. Shocked at how terrible the air in their home could be, they then bought air purifiers for almost every room.

Though the inversions aren't directly caused by climate change, the emissions that contribute to our warming world are the same that lead to our winter and summer pollution. Weather patterns that will cause more precipitation in the form of rain rather than snow have the same root problem as inversion-caused heart attacks and disruptive cognitive development in infants: toxic emissions. The dust that integrates into the snowpack, causing it to melt earlier and disrupt the seasonal water cycle, might be from the same pollution source as the dust that embeds itself in our lungs, causing respiratory illnesses. Our bodies are not so different from the snow, sensitive to the same enemies.

Not quite ten years after that murky January, I was walking across another campus, the University of Utah's, squinting through the haze in the direction I knew Mount Olympus stood. The thin, soft mask that covered the lower half of my face made breathing difficult, and the moisture from my breath made the inside of the mask wet. My eyelashes were moist from the warm air that escaped the mask, and I knew my mascara had to be running.

A woman was jogging toward me along the sidewalk, pushing a stroller. She was wearing a hat and gloves, as was her sleeping child, who was bundled up in a down jacket that puffed out around the stroller's seatbelt, but neither of them had a mask on. I couldn't help but stare as she passed; it didn't seem possible that a parent would take themselves and their child out to exercise on a "red air day," when the news stations were warning against spending time outside.

I tried not to be too judgmental; I knew I had no right to criticize a mother when I had no parenting experience of my own. Perhaps she had been unaware of the dangers of the inversion to herself and her infant. Or maybe this was her only opportunity to get any form of exercise during her day, and decided that jogging in the inversion was better than not jogging at all. In the next hour I saw at least half a dozen others exercising without a mask, and many more walking without one. Their bare faces seemed vulnerable to me, or perhaps my mask just gave me a false sense of protection.

Gathering

A gust of air froze the exposed skin on my face as I opened the car door. The coldest time of the day is just after dawn, when the sun has been hidden from a specific part of the globe for the longest, and the first rays aren't strong enough to dissolve the chilled air. Zipping up my jacket, I reasoned that though the temperature was in the single digits, it would only get warmer from here.

I was nervous. We were running a little late, and if we missed the first tram my entire plan would be ruined. Not that it wasn't already ruined; most of the plan had been shredded up last night when I got the call from Colin's dad, Randy, that the snowcat broke. I'm terrible at pulling off surprises, and the fact that I had been *so close* to pulling this one off, only to be thwarted by broken machinery the night before, hurt.

The week before, Colin had turned thirty years old. He was in Mammoth Lakes, California, for a work trip on his birthday. To make up for it, I planned a snowcat skiing adventure into a backcountry area called Mary Ellen Gulch. One of his uncles works at Snowbird as a backcountry snowcat driver, and he arranged a full day for us. I convinced Randy and one of Colin's friends from home, Michael, to fly down from Washington, an aunt and uncle to drive in from Colorado, another uncle to drive from Park City, and my own parents to commit. I even contacted Colin's boss so he could get the day off work.

In the week leading up to the surprise, Snowbird received over two feet of fresh snow. The forecast predicted a gorgeous, sunny day. Everything had worked out so perfectly that I should've known something would go wrong.

The resort was accommodating; since the cat broke, we were invited to catch the first tram and ski Mineral Basin for an hour before the resort opened to the public. Colin could tell something was wrong the night before and thought my bad mood was because of something he did. I couldn't come up with an excuse for my disappointment that wouldn't give away what had happened, so I let him continue to think that.

"Hey, Colin, what are you doing up this early?" his uncle teased, pulling skis out of his car. Colin bantered back, looking slightly bewildered to see family members from another state parked right next to us this dawn.

The rest of our group was already on the tram deck, about to board alongside ski patrol. My parents produced plastic cups and a bottle of champagne, which Colin popped as the tram swung into motion. The radiant light turned the surrounding peaks the shades of a furnace, the warmth of the hues magnificently misleading in the frigid morning. One of Colin's uncles patted me on the shoulder, congratulating me on planning the day and reuniting the Bounous and Gaylord families this morning.

Toasting to Colin's birthday, we all took sips of the champagne. The bubbles created little sparks in my mouth, reminding me of a phrase my grandmother Suzanne used to say whenever she'd drink champagne: "It sparkles like a man."

As everyone conversed, I tried to ignore the disappointment within me, hoping that no one would mention the original plan so Colin wouldn't feel the disappointment as well. Skiing on groomed runs is thrilling and fun, but it doesn't quite compare to powder skiing, when the hill isn't funneled into a manicured run but is wide open with possibilities, wild and untamed.

There's a lightness and delicateness to powder skiing, the closest you can get to flying without leaving the ground, more spiritual and out-of-body than groomed skiing. My mother once compared skiing powder to writing your signature on a landscape in a temporary, fleeting way.

The sun had risen properly by the time we reached the summit. Hidden Peak was blindingly white. Mineral Basin is south-facing, so the sun hits it hours before reaching the base of Little Cottonwood Canyon. We angled our skis, most of them sharp racing skis rather than the fat ones we would have chosen for powder skiing, down the perfectly groomed cat track. My dad and Colin's racer uncle led the pack. I leaned forward in my boots, trying to stay on their tails, determined to leave the impression on Colin's family that I was a strong skier.

The corduroy was perfect. Our edges cut curving shapes and patterns into the snow as we tore down the mountain, pausing briefly at the bottom to jeer at the broken cat that was supposed to take us out of this basin and into the neighboring one. We traversed out to Sunday Saddle, a shoulder where we could peek into the untouched gully of Mary Ellen Gulch. The Gaylords made a couple of jokes about the name; their mother, Colin's grandmother who was involved in the development of the Cliff Lodge, was named Mary Ellen.

This area, though backcountry now, has been proposed for development. Snowbird owns mining claims scattered around Mary Ellen Gulch, making much of it private property. Management hopes to expand the resort, adding 500 acres to their existing 2,500. When the idea was first introduced, my grandfather came to our house. He and my father spread out a map of the proposed area on the dining room table, bending over it together. Baqui, his dark, olive hands speckled with sunspots, pointed to peaks and ridgelines, pointing out where the lifts would start and end, tracing the boundary of the resort expansion. Snowbird was considering a gondola that would connect

American Fork, a canyon south of the Salt Lake Valley near Provo, to Snowbird, allowing residents of Utah Valley to access Snowbird without having to drive an hour north.

When Snowbird went public with the proposal they were met with outrage from many communities. Mary Ellen Gulch is a popular spot for backcountry skiing and snowmobiling, and development would mean closing it to the public. My father was asked to attend community meetings and speak in favor of the expansion—not necessarily because he supported it but because Snowbird is his employer—and he'd return from these meetings to report that those against expanding greatly outnumbered those promoting the expansion.

Another concern was water quality. Mary Ellen Gulch was mined extensively for silver and gold during the 1800s, and still has tailings scattered around the basin. Environmental groups are worried that the development of the area will disturb these tailings, causing the toxins to bleed into the watershed. Colin's cat-driving uncle, Preston, has been part of a team that has been monitoring the water quality in Mary Ellen Gulch. Their goal is to determine baseline water quality, so if development begins they'll know how it affects the water and be able to mitigate the effects.

Others who oppose the expansion want to keep the canyons as untouched as possible, believing that resorts have already impacted enough of the Wasatch. Habitat fragmentation, destructive construction, watershed concerns, and higher amounts of traffic in parts of the Wasatch that haven't been developed are some of the main concerns brought up by groups like Save Our Canyons, who typically lead the fight against resort expansion. Like many environmental fights, the battle between conservation groups and ski resorts can seem endless.

The last I've heard concerning the expansion was from my mother. At a recent owners meeting, Snowbird management said that the expansion had been put on hold, perhaps indefinitely, due to how crowded the canyon has become. On weekends parking lots fill up within two hours of the resort opening. Some ski areas, like Alta, even cut off how many people can purchase tickets. Big Cottonwood Canyon has started closing the canyon to all uphill traffic besides buses and residential vehicles when the parking lots fill up on the weekends, preventing people from parking on the street. If a storm rolls in, even weekdays draw crowds, many enthusiasts skipping work and obligations to appreciate the fresh snowfall. I've sat in lines of traffic for two or three hours just trying to get up and down the canyons on random Tuesdays—a drive that should only take fifteen minutes from Snowbird to the base of the canyon.

A common phrase heard in the outdoor community in Salt Lake City is that the Wasatch are being "over-loved." Resort expansion would allow more tickets to be sold on a daily basis, and more recreationists would be encouraged to drive up the canyon, leading to more traffic, often with just one person per vehicle. Little Cottonwood Canyon simply cannot sustain more mountain consumers. Recently Alta has taken a different direction within their business regarding the environment; though Alta has typically been seen as the more sustainably-minded of the two Little Cottonwood Canyon resorts, management is determined to expand into a backcountry area called Grizzly Gulch, despite protests from their patrons, the public, and environmental groups, including Save Our Canyons. While Snowbird appears self-aware and is changing their goals according to the environmental pressures on the mountain, Alta seemingly has decided to scrap their concerns for the watershed and the pristine solitude of the nearby backcountry.

My legs were burning during our last run of the morning, trying to make the edges of my skis carve so deeply into the snow that they'd create ripples in their wake. Colin was in front of me, his lithe body leaning over his skis as he turned. My skis crossed over his tracks again and again, creating figure eights through the firm snow. The run flattened out and we stopped, my heart rate as elevated as if I had just run a mile. As our families skied up around us, Colin took a few awkward steps toward me and put his arms around my shoulders.

"Thank you, this was amazing."

I smiled, knowing that though this day hadn't quite turned out as I hoped it would, it hadn't been wasted.

"You're welcome. Happy birthday."

"Your birthday was two days before mine—we should've been celebrating yours, too."

"You're both Aquarius?" Colin's uncle asked. "What does that mean? Do you hate each other's guts or are you soul mates? What is an Aquarius, water bearer?"

I've never been one for astrology. Neither has Colin. But as we stood there, panting in the crisp mountain air, something felt right about being called "water bearers."

Retreating

It is a frigid February night in 2002. We are dressed in full ski outfits, our jackets and pants bulging from the layers and layers of long underwear beneath them, gloves on our hands and hats on our heads. Even with all the layering, the cold bites my bones as we stand there on the street, waiting, watching anxiously.

There is a crisp anticipation in the air. The crowd around us shifts restlessly, but it's not a negative energy that runs through the masses. Everyone is fidgeting, in part to keep the blood moving through their veins, but also from the excitement that vibrates through the night. There are camera crews everywhere. Event volunteers stand at the fences that separate the street from the sidewalks, where the crowds wait.

A distant sound starts from somewhere down the street, a pitchless buzz that quickly becomes a roar. It is cheering.

My mother, aunt, and uncle, sensing what's about to happen, push my sister, my cousin James, and me up to the fence. My other cousins, Evalynn and Christine, older and taller, stand just behind us, bouncing up and down on the balls of their feet to get a better view.

A lumbering, pushing movement comes toward us, and we start cheering with the rest of the crowd. Camera lights flash through the night, illuminating the tiny snowflakes that hang in the air.

The vehicles leading the movement pass us, and we see them: two figures, one dressed in blue, the other in white, running side by side. Their faces are almost startling to look at as the camera lights flash, but we can tell who they are. In between them, raised above their heads, is a large, silver torch, a set of five overlapping rings engraved on it, a golden flame flickering at its tip.

Then it's over. The cheering moves down the street as the procession continues on.

Afterward, my mother says that that was the largest, most genuine smile she's ever seen on my father's face.

The 2002 Winter Olympics created equal parts excitement and nervousness within Utah. It was the first Winter Olympics to be hosted in Salt Lake City; my dad had joined in the effort to make this a reality. But in the wake of September 11, 2001, the final months leading up to the Olympics contained more than a trace of apprehension as higher security measures were taken within the Salt Lake Valley.

Though my father had hoped to see Snowbird host some of the alpine events, other resorts were chosen for security reasons. Little Cottonwood Canyon only has one way in, and, more importantly, only one way out. From a security standpoint, especially after the nation's most jarring terrorist attacks, Little Cottonwood Canyon was not an ideal location to host any events.

My father and grandfather were crestfallen with the decision, but were excited nonetheless to help bring one of the largest winter sports events to their local ski community. Though perhaps it shouldn't have, it came as a surprise when they were both voted to be final torch bearers in the week leading up to the opening ceremony. So on that crisp February night, my family bundled up and drove down to Provo to watch as my grandfather and father carried the Olympic torch together, as torch bearer and assistant torch bearer, as father and son.

Climate change is threatening the Winter Olympics. Both Vancouver and Sochi had to resort to extreme measures to get the snow needed to successfully pull off the skiing and snowboarding events. During the 2010 Winter Olympics in Vancouver, temperatures were too warm for enough artificial snow to be made. The event organizers had to get creative, and ended up mixing bales of hay with real and artificial snow to cover the slopes. Four years later in Sochi, organizers, predicting warm temperatures, stored snow from past years in preparation.

Of the twenty-one locations that have hosted the Winter Olympics, only a few may be fit to host them again by the end of this century. Salt Lake City will likely be one of the few cities left to host, though the venues would probably have to change from the lower-elevation resorts of Park City and Canyons to the resorts up Little and Big Cottonwood Canyons.

It's difficult to know exactly how climate change will impact the ski industry and other sports reliant on snow. If carbon emissions aren't curbed, the average ski season in North America could be cut by 32 percent by midcentury. By the end of the century, that could become 90 percent at some resorts. Because of the high elevations of the Wasatch Front, Snowbird will likely be one of the last to lose snow, but will still be affected by a warming climate: rising temperatures, precipitation falling as rain, and—due to other resorts closing—more crowds that wear down the snowpack. For many lower-elevation resorts, like those on the East Coast, precipitation will begin falling as rain and seasons will be cut too short to be economically profitable. There are plenty of resorts already that can only open for certain years, when there is enough natural snowfall to make skiing and lift operations possible. Many of the smaller, mom-and-pop resorts of the East Coast have had to shut down their operations, perhaps permanently. Resorts dependent on receding glaciers will become unsuitable for skiing. On mountains with high enough elevations to still get consistent days below freezing, snow may continue to fall for some time, and snowmaking will help subsidize storms that may not fulfill the needs of the resort.

In Alaska and Canada, climate change is threatening the Iditarod, one of the most iconic winter races of all time. Dogsled teams race each other one thousand miles through the winter landscapes of the far north, tracing historic routes that connected coastal and inland towns. The race doesn't follow an exact path, but relies on frozen waterways, like creeks, rivers, lakes, and ice in the Bering Sea, to provide passage for the sleds. Now more permanent routes, like bridges, have to be built to make up for the fact that many of these warming waterways are too dangerous to cross. In the past decade, the race has had to undergo significant adjustments to account for climate change. Twice it's had to move locations entirely, racing further north. The starting location has shifted, the race has been shortened, and piles of snow have been trucked into certain locations, providing a ribbon of white on otherwise bare ground for the sleds to travel across. Rather than snow and freezing temperatures that hold the ice in place on rivers and lakes, rain and above-freezing temperatures are changing the historical course of the Iditarod. Like other places near the poles, warming is occurring in Alaska faster than in other parts of the globe. According to the National Oceanic and Atmospheric Administration's National Centers for Environmental Information, Alaska's average temperature is warming by over five degrees Fahrenheit per century—faster than any other US state.

All around the world, climate change is shifting shorelines. The shorelines of the oceans are disappearing with sea-level rise as the shorelines of sea ice and glaciers retreat. Lakes and rivers around the world are receding as they dry up. The shoreline of snow is not as easily noticeable and harder to comprehend, but it's rising in elevation and latitude as the earth warms up. Snow is falling less at lower elevations and lower latitudes, and soon the shoreline of snow will rise enough that there will be only a few locations with adequate snow to support a ski community. These higher-elevation mountains and higher-latitude landscapes will be islands of snow during the winter, stranded by climate change. The Wasatch Mountains might have some of the final resorts remaining.

Bracing

It was a deep snow day sometime in March, and Colin and I had an amazing few hours skiing together. Halfway down our last run he asked me to take a slow-motion video of him slashing through an untracked patch of snow. When we found one, something went wrong when I tried to start the video and it only captured the last second of his turn. He got annoyed over the lost shot, so we tried again. This time he messed up his turn, and by our third try I was annoyed.

We ski differently when there's a camera around. We stop at the top of a run not to enjoy skiing it, but so that we can scope out what might look best from a certain angle. For me specifically, as the camerawoman, this means picking my way down the hill, trying not to mess up any potential powder patches, making wide and slow turns to gauge where I should put my body in regard to his and the hill. Suddenly, my last run of the day was not so much a run as it was a tedious photo shoot.

On the third attempt for the video, I situated myself beneath a tree with a rather large patch of powder at its base. He had to choose which way to come around the tree, and from my angle either approach looked good. But as he made his turn, he hit something under the snow, a log or branch that sent him off balance. He slashed through the powder dramatically and almost took me out, knocking my poles down the hill and spraying both me and his phone with a wave of snow. He blamed me that he hit the log. I blamed him that he hadn't seen it on his own.

Our silence continued all the way down the canyon, only breaking when I dropped him off at his house. The bliss that we experienced throughout the day together had deflated during the final moments of our last run.

Before I left, we watched the clip from our third try. In slow motion, his body contracts as he enters into the turn around the tree, jerking as he hits the log. He enters the patch of powder listing sideways, pointing dangerously at the camera. Abruptly he cuts his turn off as best he can, attempting to use his misdirected momentum to take him toward the viewer's left.

Slowly, so slowly, the untouched, smooth surface of the snow in front of the camera ripples, cracking and breaking into smaller and smaller cottage-cheese chunks as they're lifted into the air and toward the viewer. For a beautiful, silent moment, the viewer is surrounded by these clouds, inhabiting a sacred space between skier, mountain, and snow. It's a moment that I imagine a surfer caught in the curl of a wave would experience. Then the snow closes in, smothering everything in white.

Dolores LaChapelle, a local ski figure who was close to my family during the mid-1900s, believed that once a camera entered into a moment, "the snow is no longer a gift from the sky but a medium for making a good film." In her book *Deep Powder Snow*, LaChapelle describes skiing for a photoshoot with Warren Miller.

We took off, floating down in the powder but when we got to the bottom there was no bliss! We looked back up and our tracks were just as good as on the first run; but it was a totally different feeling. As I thought it over on the lift, later, I realized it was because I had to think about where to turn, and that's what cut me off from the on-going interactions of snow and gravity. We found that this happened whenever we had to ski for a photographer and turn in specific places. No matter how good the snow was we lost the bliss.

I've found that I enjoy a ski day less if I'm concerned with trying to get photos or videos. Entire runs can be dedicated to getting footage, taking away from the enjoyment of skiing. More often than not, the pictures don't compare to the experience. Photos and videos of skiing can be gorgeous, exciting to watch, and fun to look back on. Most of my screen savers are pictures I've taken while skiing, and I post pictures I take to my social media pages. But footage on social media outlets can be deceiving.

When my grandfather was teaching others how to ski there were not many cameras on the slopes. One of the first people to bring cameras to the hills was Warren Miller, one of the original ski videographers who made skiing and mountain landscapes accessible to a larger public. My grandpa and father were close friends with Miller. He used them in his multiple films, introducing them as "Junior" and "Junior Junior." In one of his more iconic films, *The Color of Skiing*, Miller features my grandpa skiing in slow motion to Antonio Vivaldi's "Winter." My grandpa cuts through the powder on his stiff wooden skis, white smoke billowing around him as he turns down the fall-line at the camera.

I wonder sometimes how influential that scene was in developing today's camera culture. I've had people recognize my last name in faraway places like Washington, DC, and New Zealand.

It's not just those from my grandpa's generation either; while visiting a friend in Boston, a college student connected me to the star of Warren Miller films the moment he heard my full name.

But the way Baqui places his skis on the snow one after the other is as tender as you would put a child into a crib. He swings his arms gently as he pole plants, as if he were dancing with gravity. In his aging voice are layers of love when he says the word *snow*. He says it so slowly.

On a different day in the same winter, we reached the top of the Wildcat chairlift at Alta to a strange sight. In front of us was a small hill of untracked powder and a man with a large, expensive-looking camera kneeling in the middle of it. The only reason the patch of snow was still untracked on that weekend powder day was because it required a short hike to access the hill, and because the untracked patch was barely enough to make one turn.

As we got off of the lift, we noticed that a group of four had made that hike. One, wearing a bright orange coat, pushed himself into the untouched snow and initiated a turn, rather clumsily since he didn't have much momentum. At the crux he sort of bounced, applying pressure to the snow as hard as he could with his skis, forcing a wave of snow to barrel up to his chest and head. From the lower angle of the cameraman, it would appear as though the snow was deep enough to cover the skier completely.

A few hours later a picture of a man appeared on Alta's Instagram account, orange jacket barely visible through the snow, with a caption that said the skier had gotten "over-the-head pow off Wildcat."

The word *hype*, when used as a verb, is defined by the Free Dictionary: "to create interest in by flamboyant or dramatic methods; promote or publicize showily." As a noun, it is:

"a swindle, deception, or trick." It's a common word in ski communities, often used to describe skiers anticipating an approaching storm, and getting overly excited at the thought of skiing fresh powder. Hype can also refer to the dramatic methods that resorts use to promote this over-excitedness in the community. Though they do it to encourage business, hyping is deceiving when used as a marketing tool.

The economic profit of ski resorts depends on weather patterns. Their marketing is based on getting the most out of every storm possible, since storms are predictable to a certain extent but aren't guaranteed. Snowbird and Alta, especially, have reputations based on deep-powder skiing. Resorts that want to capitalize on these storms need to create hype to attract potential consumers, hence the over-dramatic pictures of skiers disappearing into clouds of white. Ski resorts are reliant upon one real thing besides consumers: snow. And since consumers will come or leave depending on the snow, it could be said that ski resorts really are just dependent on snow.

Most resorts have not been the right types of leaders in this fight against climate change. In the weeks leading up to Snowbird's opening/closing week in November of 2017, when they opened just to close three days later, Snowbird's Instagram account was flooded with videos of powder skiing. Their posts confused the public enough that Snowbird wrote this caption under one of their powder photos: "Some people are wondering why we're posting powder videos; when you're on the hype train, you plan what you're going to do when you reach your destination, not what it's like to get off now. Keep your eyes on the prize, friends."

That November, Snowbird focused only on the potential for a good snow year, while ignoring "what it's like to get off now"— the realities of a ski resort unable to open because of a lack of snow. Snowbird could have used their platform to discuss the impact of climate change on the industry, but instead they chose to board the hype train. The following season was the worst start the resort has had since opening, besides the season of 1976/77.

Ski resorts should be at the forefront of the climate movement. They *need* to be. They should be lobbying politicians, encouraging people to divest from fossil fuels, and organizing rallies and actions for the sake of snow.

Many resorts, including Snowbird, are trying to reduce their carbon footprints. Some, like Aspen, are even lobbying Congress for climate action. But one of the most challenging aspects of climate change is public denial, and, perhaps most of all, public disinterest. Even if they believe climate change exists, many don't think that it will affect them, and don't make an effort to do anything about it. Others blatantly deny that it's happening. Colin once heard a group of ski patrollers at Snowbird discussing how climate change is a hoax with mountain guests.

No matter what ski resorts do behind closed doors to combat climate change, if the majority of their clients apparently take no interest in preserving the very thing they enjoy, I don't think the resort can claim that they're making a difference. Combating climate change has to be more than just abiding by Forest Service development standards and encouraging carpooling; it has to be a very public part of a resort's identity, saturating their business plan, apparent in the conversations between guests and employees, visible at every turn.

The National Ski Areas Association (NSAA) encourages resorts to cut back on their individual emissions with a voluntary program called "The Climate Challenge," yet in the last election cycle NSAA donated money through their PAC to a Utah politician who believes climate change is a hoax. That politician got reelected. Snowbird, Alta, Brighton, Solitude, Deer Valley, and Vail all gave money to Utah governor Gary Herbert, whose nickname is "Dirty Herbert" because of his lack of action regarding Salt Lake City's inversion. He also got reelected. The CEO of Mammoth Mountain gave over $20,000 to another climate denier, even though the snowpack of the Sierra Nevada Mountain Range will be one of the hardest hit from climate change. Jackson Hole Mountain Resort claims that they are a leader in the climate fight, yet the owner of Jackson Hole donated almost $150,000 to the Republican National Committee, and the president of Jackson Hole has sent checks to Republican politicians who support coal ports and the fossil fuel industries. A resort can claim they're taking initiative to fight against climate change, but where their money is going says otherwise.

Businesses can have strongholds on our politics. Organizations like the NRA, which has donated over $4.1 million to members of Congress since 1998, might very well contribute to the lack of action from Congress after every mass shooting. Companies like ExxonMobil and Shell lobby politicians so they won't vote to mitigate climate change. Ski resorts often get heat for bringing politics into their business models, since resorts are a place of "leisure," to "escape" things like politics. But the fact of the matter is that if the ski industry doesn't take enormous steps in the upcoming years to mitigate climate change, their industry will fall apart. Ski areas are already diversifying their business, drawing consumers during the summer months by installing zip lines or building mountain bike trails to help them stay afloat.

Some resorts are educating their customers better than others. Aspen Snowmass started a campaign in 2018 called "Give a Flake." The campaign is centered around climate change, but not from a "carpool and recycle and things will get better" standpoint. The idea behind Give a Flake is that caring isn't enough, that action must be taken. On their website, they provide resources to contact political representatives, targeting members of Congress who have acknowledged climate change but haven't taken any steps to fight against it. Aspen has also donated over $80,000 to politicians who support pro-climate legislature. And POWDR Corp, the company that owns Killington,

Copper Mountain, and Mount Bachelor Village Resorts, has given $250,000 to climate change advocates since 2010.

Imagine if other resorts, in the United States and around the world, used their wealth and rallied their patrons to support organizations and politicians who promised to move away from fossil fuels; imagine a worldwide ski industry committed to fighting climate change.

Dancing

The sound of the helicopter's rotor blades was deafening. They whipped through the air in a blurred frenzy, distorting the immediate world around them with their power. My parents were used to this alarming experience. They shielded their faces from the snow flurries the wind kicked up, patiently waiting for the chopper to take flight again. After a moment, the legs that had been resting on the snow next to them lifted off the ground in a surprisingly delicate motion. The wind from the chopper became more violent for a breath, then faded as the machine soared above them, the *whop* of the blades diminishing with its flight. My parents gathered their skis off the snow and began to prepare for their run.

It was July of 1985. A few months earlier, my parents had stood in the Cliff Lodge at Snowbird, the same lodge Colin's grandparents built in the 1970s, and read their vows to each other. It had snowed the day of their April wedding; the morning after, my grandfather arranged a heli-ski picnic in a remote location at Snowbird for the newlyweds and their families. Complete with caviar and shots of Stoli vodka lining a snow bar, it was both extravagant and wildly appropriate. It signified the joining of two families who had helped create the ski resort at which the celebration was held. The guests were flown into the picnic via helicopter while my parents hiked to a spot above the location, carrying their skis on their shoulders. Once the guests were all present, my parents made their debut as husband and wife by skiing into the picnic, their ski tracks creating figure eights in the snow behind them.

They never had a real honeymoon. A few years before their marriage my parents started a business called "Bounous International." They took clients to the South Pacific to spend a week heli-skiing in New Zealand, then to Fiji, where my mother's family owned property on an island called Vanua Levu. My sister and I are namesakes of these locations—Tyndall was named after a glacier in New Zealand, and I was named after a small, volcanic island in Fiji with just three palm trees on it, which the locals called "Asha" (my parents were creative with the spelling). Their lives had turned into one big honeymoon. After their wedding they headed back down to New Zealand in preparation for one of these ski camps, and, a week before their clients were scheduled to arrive, hunched beneath the beating wings of a helicopter at the top of a glacier.

Unlike most heli-ski runs, when the skiers are the only ones on the mountain, there were others staggered down the pitch beneath them and a small crowd at the bottom. My parents were competitors in a "Powder 8" competition on the South Island of New Zealand. This competition judges two skiers on their ability to ski down a pitch together, creating a DNA-spiral of linked "8s" down a mountainside. To score high, the skiers need to sync the rhythm of their turns with each other, one in the lead, the other just a turn behind them. The shape of their ski tracks should be round and symmetrical, their path a straight line down the slope.

The qualifying round took place on a mountain known as Black Peak. The beginning of the pitch was gradual, partially flattening out before coming over into the final, steep pitch above the crowd. My father, in the lead, picked a pace he knew my mother could follow. Their shadows danced across the snow to their side, perfectly in sync. One of the judges gave my father a thumbs-up as they flew into the flat before the steep. They came over the lip into the final stretch and my father picked up more speed than my mother could carry, his weight and strength surpassing hers. They received a nearly perfect score on the upper section but lost points for the lower when my mother got a few turns behind. But most of the other pairs had struggled in the steep as well. During their next run my mother led, setting a slower pace down the pitch, and they were selected to continue to the finals.

The final run took place on the Tyndall Glacier, the same my sister is named after. Having learned from their mistakes, they won the whole competition. My mother didn't realize she was the first woman to win a Powder 8 competition until a female-run New Zealand magazine interviewed her the following day. The qualities required in marriage—communication, trust, and a certain nuanced dance between the two participants—were the same qualities that helped them win.

My parents led these ski camps for six years, until they became pregnant with me. They skied in uncharted territory; heli-skiing had taken ahold in the Canadian Rockies and continental United States, but it wasn't really a thing in New Zealand yet. Most of the peaks and glaciers on the South Island, more mountainous than the North Island, hadn't even been skied on, and some of the only people to own helicopters were shepherds who used them to round up sheep. So my parents contacted local shepherds who lived near the mountains, mainly around a small farming community called Wanaka, and hired farmers to take their clients onto glaciers.

The isolation and terrain of heli-skiing is exhilarating. I'd say it's a nearly incomparable experience. Yet the heli-ski industry, like the rest of the snow sports industry, has negative environmental consequences. Such easy access into the backcountry disrupts local wildlife populations and releases more carbon into the air. New Zealand's glaciers went through a brief era of expansion during the last decade of the twentieth century, but their glaciers are now some of the fastest shrinking in the world. As the sea levels rise around the island I'm named after in Fiji, the glacier my sister is named after is receding. Both of our namesakes are threatened by climate change, threatened by our own family's participation in the ski industry.

Twenty years after my parents' final Bounous International ski camp, I studied abroad in New Zealand and my dad met me in Wanaka for my spring break. I was finally going to heli-ski in the place he and my mom had fallen so in love with. We walked into the office of a heli-ski company, on the edge of Lake Wanaka, to schedule a day of skiing. Upon reading our forms, the guide looked up at us sharply.

"Bounous? Your name's Bounous?"

My father nodded.

"You know, we have a run named after you here."

He took us outside, pointing to a peak that we could see clearly across the lake, its reflection mirrored in the crystalline waters before us.

Slipping

The lift swung us into the air and my grandpa lowered the safety bar. We were riding on the slow, two-person chair at Snowbird called Wilbere. My heartbeat was raised in excitement. I loved getting to sit on lifts with Baqui. It was one of the best times to ask him questions about the mountains or snow, pick his brain while I had a few minutes alone with him. The hardest part was deciding which question to ask. He speaks so slowly that he usually only gets around to answering one question during a lift.

"Baqui?" I asked, once I decided. "Have you ever seen a bear in the Wasatch?"

The Wasatch used to be home to many more bears than there are today. I have never seen a bear in the Wasatch, and sightings of them are rare. Today there are only black bears, though there

157

used to be grizzlies. The last grizzly in Utah, called Old Ephraim, was shot and killed in Logan in 1923.

"Well, yes," he responded. "I shot a bear in the Wasatch."

My excitement deflated rapidly. He *shot* a *bear*? I must've misheard him. I knew he had been a hunter, but my understanding was that he hunted for food.

"Was it threatening you?" I asked after a moment trying to find my voice. "Did it come down to your farm?"

He shook his yellow-helmeted head.

"No. I was up Provo Canyon hunting deer with some fellows. We were on an animal trail of sorts and saw it crossing a field on a hill above us. So we started shooting at it. Back then, if you saw a bear you shot it."

I didn't want to believe what I was hearing. My grandpa was so tender, so kind. I couldn't believe that he would start shooting at a bear just because it was there.

"Well, we spooked it with the gunshots but none of us hit it. It ran across the field and into the forest. We thought that was the end of it. But then, by golly, it comes barreling out of the trees and down the trail we were standing on, heading right towards us! So we all started shooting again, and one of the bullets found its mark. The bear tumbled not thirty yards from us."

"So maybe you didn't shoot it?" I asked, trying to grasp at anything to keep my childhood image of my grandfather from falling apart as we sat on the chairlift.

"No, it was my bullet that killed it," he said. "We could tell from the type I used. So I brought it home and skinned it. We still have the rug in our basement."

I was listening for a trace of remorse in his words, hoping for one. I didn't hear any. I had always likened my grandpa to Aldo Leopold, a leader of the land conservation movement. Leopold spent a period of time in his life killing predators to enhance the deer population for hunters. After killing a wolf and watching the light die in her eyes, his stance on wilderness and human's place in it changed. He developed a "land ethic" that extended to include rivers, soils, plants, and animals as well as humans, creating a biotic community. I was hoping that even if Baqui had killed a bear, he would've had some sort of enlightenment from the experience, gathered knowledge, or developed a type of intimacy with the natural world. But that didn't seem to be the case.

We pushed ourselves off the chairlift. The man who skied beside me seemed like a different one than who had sat down with me. I know he grew up during a different era, when predators were largely seen as threats to livestock and game and not as an important part of an ecosystem. The first half of his life was even before Rachel Carson wrote *Silent Spring*, and my dad recalls my grandfather using heavy pesticides on the family farm. But I had grown up thinking of Baqui as a sort of hero of the mountains, a defender of little birds and wildflowers.

I asked my father about it later.

"You know, it was a different time back then. Bears were seen as dangers and nuisances." I had to refrain from rolling my eyes at his answer. The "it was a different time" argument can drive me crazy, especially when talking about climate change and sexual assault. I didn't want to go down that path with him right then.

"Did you like killing things?" I asked, dreading the answer.

"I went deer hunting with Dad once and never went again," he said. "It wasn't for me. I never killed anything, and I didn't like watching them die."

When I didn't respond, my father spoke again.

"He stopped hunting to kill shortly after the bear, I think. He'd still go up into the mountains on horseback or foot, but just to track animals. That's what he enjoyed most about hunting anyway. The patience, the level of awareness you needed, the keen eye. He'd hunt so he could be in the mountains, but he stopped bringing a gun with him. I know this sounds weird, but you become intimate with an animal when you hunt it. I think that's why he kept doing it; he could maintain that level of intimacy with the landscape and the animal, but he got past feeling like all hunts needed to end with a kill."

Breaking

The snow crunched underfoot as my mother and I walked across the soccer field. There was a bite to the air, colder than the sunny day appeared. My parents' dog, a caramel-colored mini Australian shepherd named Bella, trotted in front of us, her dainty paws leaving dots on top of the snow. It had stormed two days ago, and colder temperatures in the valley kept much of the snow from disappearing completely. Rays from the sun had caused enough minute melting to make the outer layer around snow crystals soften then refreeze together, like ice cubes in a freezer melting slightly then refreezing. The layer of snow then compacted, becoming firmer. The frozen bonds between the crystals were strong enough to support Bella's weight so she could walk on top of the snow, but they broke beneath our boots. The resulting crunches cut through the crisp air.

Sometimes, on weekends when the buildings are dark and quiet, my mother will take Bella for walks around my high school campus. When there's snow on the ground Bella will run in large, looping circles for a good portion of the walk. My mother invited me this morning and, a little to my own surprise, I accepted. A heaviness had plagued me since I had woken. I couldn't bring myself to go skiing. Even a mild walk around my old campus seemed a bit ambitious, but manageable. I'd be furious with myself if I wasted another beautiful day inside.

We approached the edge of the soccer field and Bella turned her head toward an undeveloped field to our left, her beetle nose twitching at the scent of something.

Probably a killdeer, I thought, remembering when my high school biology teacher used to bring us to that field. Killdeer are birds with black-and-white stripes around their necks. They build their nests in the ground in flat, scrubby areas, which the campus had a few of. When something wanders too close to one of their nests they'll feign injury, dragging a wing behind them in an attempt to lead the intruder away. My biology teacher instructed us to fan out across the area, walking toward any killdeer we saw to try to get them to fake a broken wing. I looked to see if any killdeer were moving between the sagebrush, but the field seemed as still as the campus.

"You okay Ayj? You seem a little quiet."

I tried to gather my thoughts, to organize them in a way that would make sense to her. These nihilist feelings I had were often hard to vocalize.

"I'm having a hard time imagining what a future here will be like."

The words sounded more dramatic when spoken aloud than they had in my head. The crunching beneath our feet stopped abruptly as we stepped onto the snowless sidewalk.

"Do you mean . . . a future with Colin?"

I shook my head.

"No. Well, maybe—it all seems so connected to me—" My voice caught. What *did* I mean? I took a breath, chose a direction.

"I'm frustrated with the ski community. So many of them don't seem to care about climate change, even the ones who believe in it. Even when Colin talks about climate change it's not really about how it's going to affect us personally, it's usually about how it'll impact places and people far away."

"That reminds me of a study that came out recently . . . I think Yale published it? It was about how most Americans believe in climate change but don't think it'll affect them."

I nodded. I had seen that study too. It included maps of counties, with different colors representing different levels of climate acceptance. Utah had some of the "least-worried" counties in the country, even though the American West is suffering from climate-change-induced drought.

"Less than 50 percent of the population thinks climate change will harm them," I said, "and Utah is basically the worst; Morgan County, just north of Salt Lake County, has one of the lowest percentages in the country."

"So Utah is in denial."

"Right. And I think the ski industry is one of the communities most in denial. I don't think many even realize how climate change will impact them, even if they believe in it."

Bella rummaged her nose into the snow beneath a bush, sending puffs of white powder into the air with each snorted breath. The puffs sparkled as the sun flashed off of the tiny crystals.

"I'm frustrated with them, and I'm frustrated with myself. I don't know how to change it," I continued as Bella lost interest in the bush. "How do you get a community to care? Sea-level rise is one thing; that's a tangible threat to a landscape and the people reliant on that place. And natural disasters like hurricanes and wildfires are so violent that they force communities to pay attention. But when a winter season can vary so much from year to year, it's hard to quantify and show how climate change is affecting our mountains and community."

"Perhaps it's easier to focus on saving a landscape than on saving a season," my mother responded after a moment. In the distance, a bird called *ki-deer! ki-deer!* There *had* been a killdeer hiding somewhere in the sagebrush.

Bella had started her looping laps through the snow. Her paws left patterns on the surface, tracing her progression. My friends and I used to write our names in the snow with our footprints during lunch breaks, or create massive designs of flowers, suns, or spirals. I was a big fan of making raindrops. Patches of snow I left behind often looked as though the white surface had tears dripping down it.

It was on this campus I had my first few years as an environmental crusader. In the wake of Al Gore's *An Inconvenient Truth* the "green" movement of the early 2000s took flight, calling on citizens to make changes in their everyday lives. I latched onto the movement: pulling plastic bottles out of trash bins and calling out classmates who threw away items that could be recycled, covering my shiny new hybrid car with bumper stickers that said "Save the Whales" and "Love Your Mother," turning off lights when I'd leave a bathroom, sometimes accidentally when there were others still in there. As sophomore class president, I made my grade "adopt" a polar bear named Snowflake, as though the small monetary donation could do any good in preserving the species. In short, I was obnoxious. I can't say that my tactics made anyone change their ways, believe in global warming, or even be sympathetic to the cause.

My efforts seemed downright naive once I began comprehending the true scale of climate change. As I understood that nothing I could do in my personal life would change the outcome, I swung from ambition to depression. Climate change was easier to cope with when I felt like my day-to-day actions could make a difference. My momentum deflated fully once I heard that just one hundred companies were responsible for over 70 percent of greenhouse gas emissions since 1988. I couldn't even redistribute my efforts elsewhere; I didn't know what I could possibly do as an individual to make a difference.

Rebecca Solnit, author and activist, has said that "action is impossible without hope." I know this is true. When hope leaks out of me I lose all steam, any drive to push me forward. As I kept stride with my mother, watching Bella circle back to us with her tongue hanging out, hope felt intangible and fleeting, like wisps of smoke in the air.

"You say you're having a hard time imagining a future here. Have you thought about how your thoughts of the future would change if you pictured your children in it?"

My children? I didn't understand what she meant. We had had conversations about my hesitance to start a family. My mother is a realist when it comes to climate change. She *knows* our climate is headed for worldwide disaster, but can't grasp why I would forgo having children. I couldn't keep the impatience out of my voice.

"The future is *why* I don't want children. Not starting a family is probably the only real, tangible thing I can do as an individual to lessen my personal contribution to climate change. Recycling, driving less—none of those things really matter while big oil, gas, and coal companies control our politicians."

This was the hardest part about having these conversations with my mother. There was a disconnect between us; she saw my decision to not have kids as being a choice. I did not. This decision to start a family no longer felt like my own. If I didn't want to have kids for personal reasons that would be one thing, but this choice wasn't personal; it felt like it had been taken from me. Bearing children is a choice a woman *should* be able to make. When that choice is robbed from you by others, by large companies that control the world's fossil fuel economy . . . my ears were ringing. My heart beat in my throat.

"How can I have children if I don't have hope?"

She answered my question with another question.

"But what if choosing to have a child is choosing to have hope?"

We rounded a building and came face to face with a woman on the sidewalk, pulling a small, shiny red wagon behind her. A young girl was in the wagon, giggling uncontrollably. Golden curls the color of Colin's spiraled out from under a pink beanie with a pom-pom. She was bundled in a snowsuit, tiny mittens on her hands, and I was suddenly reminded of a silent auction I attended a few years ago, when a few too many glasses of wine caused me to be the highest bidder on a child's ski suit patterned with foxes and raccoons and mountains. I kicked myself for days after the purchase, yet part of me was secretly thrilled when I pictured my future child in it.

I had always imagined that any children I had would look like me, with darker hair and grayish-blue eyes. But as the young girl giggled up at me, her eyes bright and her curls glistening like ribbons of sun, a yearning to have a child who looked like Colin filled me with such intensity that for a moment I forgot how to breathe.

Floating

The quiet between the trees seemed almost too silent to be real. We stood for a moment, hesitant to make a sound, as if the trees surrounding us would stir with disapproval if we disturbed their peace.

There are many things that can seem mystical about a landscape freshly covered in snow. The way the world softly glows at night is one. Another is the way that snow absorbs sound. Snow doesn't just soften edges, corners, and colors; it softens sound.

The space between crystals, which keeps fresh snow light, fluffy, and fun to ski on, is also soundproofing. When sound travels through snow, rather than bouncing off the surface and traveling further like sound waves do with other materials, the waves disappear into the porous gaps between crystals, muffling noise. As snow melts and compresses, or turns to ice, this quality changes. A hard, icy surface will reflect sound waves, making noises louder in frozen landscapes.

Colin was the first to break the silence, knocking the thick snow that had accumulated on the bottom of his boots off with his pole. The canopy of the surrounding pines here was higher than I was used to, the trees more spaced out than most runs at Snowbird, where heavy logging in the 1800s removed most of the old-growth forest that used to span the Wasatch. But this place felt like a small nook where old growth remained, where ancient pines with twisting trunks still dominated the hillsides. Identifying pines had never been my best skill, but I could still recognize the diversity here; some had bark so gnarled and gray they looked like they could be the legs of ogres or giants, others were smoother and more graceful, with lighter hues of mahogany and sepia. And between the trees, so soft and bottomless it was as though we were skiing through cotton, was some of the most magical snow I had ever encountered.

We were skiing at Solitude that day, one of the resorts up Big Cottonwood Canyon. It had snowed twenty-eight inches in the last twenty-four hours, so deep that my thinner skis were having trouble staying afloat, especially on the flatter terrain. The second I'd begin losing momentum Colin would sail past me, his longer, fatter skis charging through the lower-angled pitches.

I wasn't complaining, though. Of all my days, I had rarely experienced anything like this. There had been a line of cars going up the canyon, a full parking lot, a line to get on the lift, but we had found a spot on the mountain where no one else was venturing. Or perhaps there were others there, but the snow was falling so thick that their tracks would fill in before we could make a lap back to the top. Every time we finished the hike I'd try to look for our last tracks, unsuccessfully.

The route we were lapping required one hike, two traverses, navigating around a cliff band, a long run-out flat area I'd inevitably have to push myself through, and three lifts. It took about an hour to complete, and we had managed it four times. My legs were exhausted, in part because of all the hiking and sidestepping required, in part from trying to keep my ski tips afloat.

If I were to point to one ski day during my life that I'd name as the best, this one would make it to the final cut.

Colin started down the hill first, cutting right through the trees. I followed, weaving my way left of his path. The trees were spaced perfectly, the pitch steep but not too steep. Each turn was effortless. My legs, my ankles, my core were all working together to stabilize me, but I wasn't aware of any of that. There are times when skiing is hard, when you have to focus your energy and strength to move through the terrain. There are times when I am afraid, or hesitant to turn and carry speed. And then there are times like these, when it happens so fluidly, so formlessly, that you're barely aware of your body in the landscape, all sense of self flees, and you're left with just movement and exhilaration.

In her book *Wanderlust: A History of Walking*, Rebecca Solnit says that "walking . . . is how the body measures itself against the earth." If walking does this, then skiing is how the body measures itself against the earth, the sky, weather patterns, gravity, and everything in between.

Our family friend Dolores LaChapelle takes the teachings of the philosopher Martin Heidegger, who emphasized the "authenticity" of the human experience, and applies it to powder skiing. LaChapelle believes skiing the most "authentic" way to experience what Heidegger calls the "round dance of appropriation." For LaChapelle, this dance is "the interrelationship of the fourfold: earth, sky, gods, and mortals in *my* world of powder snow skiing." In the round dance of appropriation, a relationship is created between the sky and the earth, weather and mountains, the resulting "gift" being a landscape of snow. Skis become a vehicle for humans to experience this unity. LaChapelle believes that once a skier enters this relationship between earth and sky, snow and gravity, she becomes one with this process, and the entirety of this encounter is called *Being*.

> Once this rhythmic relationship to snow and gravity is established on a steep slope, there is no longer an "I" and snow and the mountain, but a continuous flowing interaction. I *know* this flowing process has no boundaries. My actions form a continuum with the actions of the snow and gravity. I cannot tell exactly where my actions end and the snow takes over, or where or when gravity takes over.

Gravity had taken over. I was carrying speed through the trees and it was effortless. Colin was beneath me in the middle of a slightly steeper pitch, looking up the hill to make sure he hadn't lost me.

In that moment, I saw my perfect line. Just to the right of him was a slight opening in the trees, and I angled toward it. My vision shifted; instead of picking out the trees my brain was picking out the negatives spaces between them, illuminating my path. I picked up speed as I dropped down the pitch, the snow billowing off my chest and into my face. I was aware of passing Colin, sensing his presence, and not much else. In that moment, my body was not mine. It bled into the landscape around me, into the snow and the trees and the earth and the atmosphere, fleeing into those spaces between crystals. And the landscape bled into those spaces within me.

I wasn't fully aware of stopping, but once I did the world came rushing back. I was panting. My legs were burning. My lips were wet with snow.

I looked back up the hill at my tracks, feeling as though I had left part of myself up there, buried beneath the trees. In return, something remained within me, filling the spaces between ribs, crystallizing around blood cells. It was not unlike the feeling of being in love.

While living in the depths of a desert, author Edward Abbey once said, "I dream of a hard and brutal mysticism in which the naked self merges with a non-human world and yet somehow survives still intact, individual, separate."

I emerged from that dream intact, but I was no longer individual, separate. My body was saturated with Wasatch water. Carved from crystals. Shaped by snow.

Ellen Meloy would call it a *glimpse*, Martin Buber would call it an *encounter*, Edward Abbey would call it a *brutal mysticism*, Thoreau would call it a moment of *contact!* Heidegger and LaChapelle would call it *Being*. I would call it *birth*.

V. Melting Season

Sprinkling

I t only took a run for me to notice that something was off. My skis were sticking to the snow on the groomed terrain, and the fresh powder between the trees was thick and hard to turn in. The snow practically squeaked under my edges as I pushed myself through the lift lines. I made a comment to Colin as we sat on the lift, and he laughed.

"This is how the snow usually is in the Pacific Northwest. Actually, this is still better than most of the skiing there. They pass out garbage bags to ride the chairlifts with at Mount Baker."

The types of snowflakes that fall during a storm depend on the temperature of the clouds that the snow crystalizes in and the temperature of the air that they fall through. Because warmer air holds more moisture than colder air does, snow that falls around or just under freezing temperatures holds more moisture than snow that falls in more frigid temperatures, so the warmer the location the higher the water content. Warmer snowfall means bigger flakes that stick together as they fall, creating a stickier surface that makes skiing more difficult but snowball-making easier, while colder conditions typically result in smaller, individual snowflakes that are more likely to be blown around by the wind once they land.

Local factors influence Utah storms and snowfall. When the winds blow from the northwest, they scoop the warmer air that hovers on the surface of the Great Salt Lake. When the moisture-heavy air gets lifted up to the cold, high elevations of the mountains, it can cause lake effect, contribute moisture to the storm, and add another inch or two of snow.

Another localized factor is the elevation difference between the valley floor and the mountain peaks—about 4,500 feet in the valley and 11,000 feet at the summit of Hidden Peak. According to atmospheric scientist Jim Steenburgh, the difference between how much snow falls at the base of Little Cottonwood Canyon versus the top is one the most significant differences in the world. In general, there's an increase of one hundred inches of snow per one thousand feet of elevation gain as one travels up the canyon, which translates to an average of one hundred inches of snowfall at the base and five hundred inches at the resorts.

The Cottonwood Canyons, especially, are known for their microclimates. The resorts up Big and Little Cottonwood Canyons benefit from the greatest variety of storms in the Wasatch. While many other resorts might experience heavy snowfall from storms that come from the southwest but light snowfall from storms that come from the northwest, or vice versa, because of the height and broadness of their peaks and ridgelines, the Cottonwood Canyons benefit from most of the storms that hit the Wasatch.

The storms themselves are remarkably consistent. When temperatures in the Wasatch are warmer, we, too, can experience the "cement" associated with the Sierras and the Cascades, though it's not very common. The trend of Wasatch snow storms is to start with warmer temperatures that decrease as the storm continues, leading to conditions that snow scientists call *right-side-up* snowfall. When the storm is warmer, it creates heavier snow. But as the storm drops in temperature the snowflakes that continue to fall become lighter, covering the heavier layer.

This vertical weight distribution in the snow creates "body," making it easier for skis and snowboards to float on top of powder. The heavy snow on bottom creates a denser layer that supports the fluffy, seemingly weightless powder on top, and the result for any recreationist is the feeling of simultaneously falling and floating down a mountain.

Right-side-up snow is also healthy for the snowpack. When *upside-down* snow occurs, when the storm warms up as it develops and drops denser snow on top of lighter snow, it not only becomes more difficult to ski but it creates unstable layers in the snowpack, more likely to create slabs that can lead to avalanches.

That morning, after checking the storm totals, I had remained in bed as Colin started to get ready. I watched anxiously as he pulled on his long underwear. It felt like a while since we'd last skied together. The snow of the past few weeks hadn't been that great, and with each big storm that came there were so many crowds that I had been avoiding making the journey up the canyon, making excuses for myself. My father had warned me that this storm would be moisture-heavy, and I knew what that meant for my inflamed lower back.

Colin glanced at me, the covers still pulled up to my chin, and asked if I was going to stay at home. I stayed silent at his question, fiddling with the sheets.

He kept his head down while he said it, pulling his ski pants on, and though his voice had a playful tone to it, his words stung.

"You never want to ski with me anymore."

Dense, wet snow creates more risk for avalanches. Earlier that season, Mount Baker closed its slopes due to avalanche danger from twelve new feet of heavy snow in ten days, topped off by a day of rain as the weather warmed up. Warmer ski areas can get massive amounts of snowfall due to the moisture in the atmosphere, but the quality of the snow that falls can be frustratingly heavy, often un-skiable, and very dangerous.

After another particularly wet storm, I went to my parents' house for dinner. I zipped off my rain boots as I entered and beelined for the open bottle of wine on the kitchen counter. I had just spent hours running errands in the rain with just one windshield wiper working. I poured myself a big glass.

"Hi, Ayj," my mother said as she opened the oven door.

"Feels like proper spring, doesn't it," I said by way of greeting. Then, registering what she was doing, I asked, "Are those gloves?"

"Yeah, we're trying to dry them out," she responded, shoving four pairs of ski gloves in the oven. I glanced around the room. Items of clothing hung off of every piece of furniture. With varying shades and textures of wool and polyester and Gore-Tex, the house looked like some sort of foreign market. My dad crouched near the dining room table, spreading more layers of wet clothes out on the tile to take advantage of the radiant heating.

"Can't you put those in the drier?" I asked.

"Tyndall's put her uniform in there," he said as he smoothed out a puffy jacket. "Turns out the jackets and pants Alta bought their ski instructors aren't waterproof."

"I couldn't even put my phone in my pocket!" my sister said indignantly, walking down the stairs. "And none of our clients could keep their paper lift tickets in their pockets either."

"It was the worst day I've ever had at work," my dad grumbled. "I went through five pairs of gloves. We started prepping the race course in the rain at five a.m. We used eighty-five bags of salt—forty-five pounds each!—to try to get the course to refreeze, but we ended up having to cancel the race. We'll have to do it all over again tomorrow. The resort couldn't open two-thirds of the terrain because of avalanche risk."

"There were a bunch of avalanches up at Alta, too," Tyndall said as she poured herself a glass of wine even larger than mine.

My dad shook his head, his face etched with disbelief. "Ski patrol set off a few inbound avalanches with fracture lines that cut all the way to the ground. You could see the dirt in the avalanche path. The rain rotted all the way through the snowpack."

In a study conducted by the Climate Impact Lab in 2018, the number of days below freezing in some locations in the Wasatch could be cut in half by the end of the twenty-first century, resulting in more rain and less snow. Because rain passes through a watershed quickly, it will disrupt the ebb and flow of the surrounding ecosystems. This will cause inconsistent water levels in rivers and streams, affecting riparian vegetation and animals, and drier soils that make the growing season tough on high alpine environments. The highest elevations will be less at risk of rain initially, but eventually the elevation required for snow will pass the elevations of the Wasatch.

It will also, of course, disrupt the skiing. As my family complained about their day in the rain, I imagined donning a plastic bag to ride the chairlifts at Snowbird, no longer home of the greatest snow on earth, imagined my sister holding an umbrella over her clients as she taught them how to ski. I pictured myself staying at home during a winter storm, deciding the snow wasn't worth it to ski with Colin.

Dissolving

I came through the mountains and descended into an apocalyptic scene. For the last three hours I'd driven beneath blue skies and the white desert sun. I noticed the wind picking up while gassing up in Scipio, but didn't think too much about it. The desert is windy. The Salt Lake Valley, however, was a churning landscape the parched colors of rust and rot. My tongue shriveled just looking at it. It was a landscape of nightmares.

The afternoon sun had dissolved into the fine dust. Winds had scoured dirt from the western deserts, pulling it over the Oquirrhs and into the Salt Lake Valley. The Wasatch Mountains had all but disappeared into the layers of silt, as hidden from the valley as they are during inversions.

I came through the mountains and into the future. Climate change predictions declare the American West as the next Oklahoma Dust Bowl of North America. A study published at the end of 2016 estimates that the likelihood of a megadrought, lasting decades, is now close to 100 percent if carbon emissions aren't curbed. Another study, released at the beginning of 2019, claims that we're already in a megadrought. With less precipitation and more soil cut and eroded by ranching, agriculture, and extraction in the Great Basin, Salt Lake City could become the Western Dust Bowl. Wind and dust will cripple the American West.

I was returning to this dire scene after a weekend getaway to southern Utah. There, cryptobiotic soil builds the desert into fertile landscapes. This soil, which grows at the slow rate of one millimeter per year, has a layer of life on its surface which holds the soil in place, preventing erosion. But once that crusty layer breaks under some sort of stress, like a footstep, it begins an erosion process that allows the wind to stir up the red dirt from the deeper layers of the landscape. Climate change is also a threat to these soils, drying them out, cracking them, opening up the heart of the American Southwest to the elements. Once the soil is damaged, it can take 50 to 250 years to heal.

From the 1860s to the early 1900s the amount of dust present in the atmosphere increased by 500 percent, due mainly to intense and unregulated cattle grazing. Today, developments, the extractive industries, roads, dried-up lake beds, and continued livestock grazing are the main sources of dust. Areas scorched by wildfires can also be major sources. Though the desert might seem like it would be a natural place for massive dust storms to originate, when a desert ecosystem is healthy, the seemingly scarce vegetation holds the soil and sand in place. When a wildfire or development eliminates or destabilizes vegetation and root systems, that area becomes a point source of dust.

Once an area has been damaged in some way, it becomes more vulnerable to wind storms. Since many of the mountains in the American West are within or border the Great Basin Desert, dust from the desert often winds up in these mountains. Though dust and wind are considered secondary effects of climate change, the impact that dust has on snow is more negative than direct atmospheric warming. Once dust is integrated into the snowpack, the snow's albedo, its ability to reflect light, declines, resulting in more light and heat absorption. Salt and certain minerals cause the freezing point of water to drop and snow crystals to melt. Because of the darker and mineral-rich layers, a snowpack integrated with dirt will melt quicker than a snowpack unaffected by it.

Mountain ranges with snowpack that has been subject to dust storms will experience spring runoff three weeks earlier than their dustless counterparts. Mountain ranges in western Colorado have already suffered from runoff that came forty to fifty days early, disrupting vegetation and life downstream and causing the soil to dry up quicker, leading to stressed summer alpine ecosystems. Even if a winter is cold or not unusually warm, if dust has been integrated into the snowpack it can still cause unnaturally early snowmelt. Runoff caused by dust is the equivalent of runoff that could be caused by two to four degrees of temperature change, and is considered more of a risk to snow than actual warming. According to a recent study about dust in the West, returning to "pre-mid-1800s dust conditions" would keep the western snowpack healthy, mitigating the direct effects of atmospheric warming from climate change.

Restoration work and regulations that aim to address dust release could limit the amount entering the atmosphere and harming the snowpack, but many of the originating locations are difficult to control. Areas affected by wildfires can be hundreds of acres, and the soil can be damaged enough that getting new vegetation to grow can take years. Areas that have been subject to development, like highways or housing developments, are continuous sources of pollution, especially if they're unfinished or abandoned. Extractive industries, like oil, gas, and mining, release dust from places of extraction and from increased traffic to and from those areas. Open pit mines, like the Bingham Canyon Mine and the multiple gravel pits found in the Salt Lake Valley, have no way to keep dirt and particulate matter on the ground, creating localized pollution. Surrounding neighborhoods can experience health hazards, like asthma and reduced visibility, that can be directly traced back to these sources. West of the Salt Lake Valley, craters created during the weapons testing of the Cold War also release dust into the atmosphere. Weapons testing continues to this day. Dust originating from those areas is subject to metal and even nuclear poisoning. My father recalls instances growing up in Provo when what appeared to be snow or white soot would fall from the sky, and his teachers would have to warn the children not to eat the nuclear fallout.

Because of rerouted rivers and continued damming, the shallow expanse of the Great Salt Lake could dry up as well, evaporating and leaving behind massive amounts of salt. The salinity could lead to crop failures anywhere downwind of the lake. Salty and radioactive dust storms, worse than a biblical locust invasion, could plague not only the crops but the soil itself, making the region unproductive far into the future. Respiratory illnesses, heart failure, and even conditions like Alzheimer's and autism have been linked to pollution from particulate matter. Contagious diseases—bacterial, viral, protozoan, or fungal—can also piggyback on dust particles, aiding the spread of transmissible illnesses. Cue the apocalyptic American West.

❄❆❄

After hours of battling the intense wind on the freeway, I entered Colin's house to find him irked. He told me that he went up to Snowbird only to find out that the tram had closed; it had hit one of the towers in the wind, and the snow conditions weren't worth the lift lines. So he got right back in his car and drove down the canyon.

"Is that it?" I asked.

No, that wasn't it. It was too windy to golf, too.

Vanishing

While growing up, a phrase I'd often hear from my mother was, "The girl who skis gets the man."

She developed that phrase from years of visiting Sun Valley, Idaho, as a child. Since it was where her parents met, the resort was as much of a family tradition for them as Snowbird and Alta are for us. There, at the Sun Valley Lodge, a black-and-white film plays on repeat. Made in the 1930s as a promotional film for the railroad that connected the East Coast to the Mountain West, *Sun Valley Serenade* has ski chases, tap dancing, ice-skating, a polka dance, broken skis, scandalous encounters in a ski hut, and music lyrics that my mom likens to my maternal grandparents' courtship in Sun Valley during the 1940s.

"It happened in Sun Valley, not so very long ago . . ."

To simplify the storyline, the film is about a man who chooses between two women. One is a beautiful singer who doesn't enjoy spending time in the out-of-doors, except in a bathing suit by the pool. The other is a refugee from Norway, an avid skier and ice-skater. In the end the man, who also loves skiing, chooses the woman who skis, hence my mother's mantra.

My mother's parents, Grant and Suzanne, met in Sun Valley. My grandfather's roommate, Suzanne's brother, invited Grant on a family ski vacation there. Suzanne had just returned from traveling with her mother for a few months in Norway, where a Norwegian man had proposed to her. The way my grandfather told it, Suzanne wore the engagement ring the first night of the trip. By the second night, the ring was nowhere to be seen. In my family's storyline, it wasn't the man choosing between two women, but the woman who chose between two men.

My parents continued the Sun Valley tradition by bringing us every year for spring break. An eight-foot-tall sun with a smiling face, carved out of ice, would greet us outside the hotel. The first thing my sister and I would do once my parents parked the car was run our soft, plushy fingers against the sun's watery lips. But each time we visit Sun Valley it seems to be warmer, like the Salt Lake Valley, no longer able to support an ice sculpture. Besides during a visit over Christmas break one season, we haven't seen that sun in years.

After a nearly five-year hiatus, my family returned to Sun Valley for an impromptu vacation. In lieu of tracing the absent sun's lips with our fingers, my sister and I helped our parents drag layer upon layer of ski bags out into the fifty-five-degree weather, the eyes of the bellmen widening with each unearthed bag. We passed through the golden revolving doors into a polished room of dark wood, golden accents, and swirling patterns of strawberry red and white.

The hotel lobby is lush and extravagant, steeped in mid-century decadence. Since making its way to the American West, skiing has been a form of recreational leisure for the wealthier classes, noticeable especially in the more isolated resort towns, like Sun Valley. Beautiful log cabins, many of them second homes, decorate the hillsides here, signs of economic prosperity. Yet on the way into town are rows upon rows of tiny trailer-park homes, a stark contrast to the wealth of the resort community.

Mountain towns experience some of the highest income inequalities in America. Jackson, Wyoming, and Aspen, Colorado, are among the cities with the highest income inequalities in their states. Ski resort workforces are largely based on immigrant communities, often paid minimum wage despite the spiked cost of living in mountain towns. The American ski industry relies upon workers born outside of the United States, whether seasonal employees who travel from Europe to work serving jobs, or through more permanent custodial jobs, often filled by immigrants from Latin America. In some counties of Colorado, up to 30 percent of the population is Hispanic, though their culture is greatly unrepresented in most resort towns. A general manager of a hotel at Mammoth Mountain once said in an interview with *POWDER Magazine* that 100 percent of its hotel employees are from Mexico, yet a tourist visiting Mammoth would likely not see any hints of Mexican influence in the town itself.

Sometimes the two communities within the ski industry can seem like the two sides of the moon, one side bright and easily noticeable, the other hidden and even ignored, yet both part of the same system. Many of these employees don't benefit from the complimentary ski passes included in their wages; most can't ski, or won't, due to risk of injury, but may spend their free time hiking or appreciating the mountains in other ways. Their children, who attend schools with designated ski days, are often given the opportunity to learn, and many love skiing just as much as their peers, participating in race teams and considering future careers in the industry.

For those of us integrated into the more "noticeable" side of the ski industry, meaning tourists or those whose jobs are more directly based around skiing, like instructors, it's easy to forget about or be unaware of others who are as reliant on healthy winters as we are. Their communities might not draw as much attention as recreationalists do, but if the winters became unreliable enough to cause them to leave mountain towns, the entire industry would be affected.

Just as my grandparents and parents decided to invest in snow and the ski industry, these individuals have invested in it as well. Unlike my grandparents—one set who had investments all over the world and the other set with a farm to supplement their income—and my parents, who had the means of choosing different lines of work, these populations are reliant on the flow of income into mountain towns not from a recreational standpoint, but to sustain their livelihoods. They need snow to bring business. I can't speak to how their relationships are shaped by winters. I cannot claim myself to be part of their community, so I won't try to hypothesize as to how their stories will change as the winters continue to warm. But if the economic flow into mountain towns runs dry with the future droughts of the American West, they will become more displaced than the ski bums and upper-class folks who live up the street.

It was late March of 2015 when we visited Sun Valley last, and the wing of the hotel we stayed in was the only one open. The Sun Valley Lodge was going through some "much-needed renovation," according to hotel management. Apparently it hadn't been remodeled since opening in 1936. They promised to keep the classic, cozy atmosphere the Lodge has welcomed guests with for the last eighty years, but a rumor that the ballroom would be destroyed reached our ears during our visit.

I tried to push the thought out of my mind until the last day of our trip, when my mother and I walked down the hall and into the ballroom for the last time. I imagined my mother as a little girl of four or five, dressed in a billowy cream nightdress, wandering down this same hallway. Fifty years ago my mother had a bad dream, and instead of running to the babysitter in the next room she ran down the hall and into the ballroom, across the wooden dance floor between the floor-length dresses and tailored pants of the dancing couples, under the glistening twinkles of the crystal chandelier overhead, and into my grandmother's lap, burrowing her buttery hair into the folds of Suzanne's sequin skirt.

The windows hadn't been washed in some time, and the dust created a veil that hung between us and the melting world outside. My mother walked onto the dance floor, her slight frame silhouetted by the afternoon light. Her shoes clicked softly on the wooden floor as she passed under the chandelier. She glanced briefly in the direction of the grand piano, where her mother used to play, and through her eyes I saw my grandmother in a deep-purple chiffon dress, her dark hair twisted elegantly on top of her head and swan-like neck as she sat at the piano, my grandfather behind her in a three-piece suit, his hands on her shoulders.

I used to imagine myself getting married in this room. After skiing, or ice-skating, I'd wander into the ballroom during the day when it was empty, and imagine flowers and ribbons draped around the banisters. I'd see myself, taller and more beautiful, perhaps resembling my grandmother, sharing my first dance with that faceless someone, dancing on the same floor my grandparents and parents have danced on countless times.

Knowing I was standing in this ballroom for the last time, I conjured these images from my imagination once more, this time the faceless someone not so faceless. The figures shimmered in the fading room like a mirage: ghosts of a disappearing future overlapping with the small figure of my mother, running across the dance floor in a billowing, cream nightdress.

Receding

Stepping out of the car, the cold air of early spring gripped my bare ankles. It was as alarming as being grabbed by the hands of a corpse. Startled, I glanced at the thermometer on my dashboard. Thirty-nine degrees! Utah had tricked me again. I couldn't say that I was surprised, but as I began packing my things into a backpack I was grateful I had brought a thin pair of gloves. I shouldered my backpack and shivered, harboring regret at not wearing full-length pants.

The reflection of the morning sun on the bright whites of the mountains had tugged me out from under my roof and into my car. I needed to see them from a distance, needed to be reminded of their beauty, their unwavering stance along the Salt Lake Valley. Anxious over the lack of moisture in our watershed, I was craving the reassurance that, though the climate might be changing, the mountains themselves weren't going anywhere.

In the distance, the water of the Great Salt Lake flashed those white peaks back at me. Though wind was feathering through the silvery fingers of sagebrush at my feet, there didn't seem to be strong enough gusts on the surface of the water to smear the reflections. The sky, a turquoise so light and airy from the high-pressure system over the valley that it seemed as though it was dissolving into the atmosphere, also reflected in the lake, captured in the land like light in a jewel.

I was on Antelope Island, a shard of high desert surrounded by the Great Salt Lake. The largest of the ten islands scattered throughout the lake, Antelope Island is home to one of the largest and oldest publicly owned bison herds. Back in the 1800s, when bison were practically extinct in North America, twelve bison were brought to the island by boat for breeding and hunting purposes. Now five hundred to seven hundred individuals live on the island, and the population is considered an important feeder for other bison herds across the continent. The texture of their massive heads against the brushstrokes of golden grass, crystal glass lake, and fractured mountain peaks in the backdrop causes traffic jams on these island roads.

The Great Salt Lake is an important yet under-acknowledged feature of the American West. The largest lake west of the Mississippi and the fourth-largest lake in the world without an outlet, it's saltier than the ocean and a sanctuary ecosystem for migratory birds. Bison might be the most traffic-stopping wildlife living near the lake, but the body of water is home to a wide array of animals dependent on its salty shorelines, including brine shrimp—locally known as "sea-monkeys," which provide critical food to bird populations—brine flies, pronghorn antelope, coyote, mule deer, badgers, bighorn sheep, bobcats, jackrabbits, gopher snakes, and ground squirrels. It's also the winter home of over five hundred bald eagles, one of the largest populations in the continental United States.

Seven-and-a-half-million birds and 257 bird species are reliant upon this ecosystem. The Great Salt Lake, smelling like rotten eggs and too salty and muddy to enjoyably swim in, might seem like a desolate space to human residents, but it's an oasis of resources amid the Great Basin Desert for birds making their yearly migrations. At times, a third of the world's population of phalaropes, shorebirds recognized by the rust-colored patch on their necks, pick their way through the marshlands. Two-and-a-half-million eared grebes, dark, duck-like birds with cinnamon-colored "ears" that swoop upward from their red eyes, also reside here—about half of the grebe population in North America.

I was hiking on the shoulder of Frary Peak, the highest point on Antelope Island. I was not near the shoreline, yet I was still playing witness to active birdlife. A nighthawk, the white line just beneath its "elbows" looking like it could've been smeared on its wing by a fingertip, cut across the light sky while I was driving. Within minutes of starting my hike, I heard a sage thrasher begin singing relentlessly from the brush to my left.

I navigated to a rocky outcrop facing the Wasatch. From my perch, the water appeared no more than a thin layer of glass, or a mirror upon a tabletop. The image is not far from the reality. The Great Salt Lake is an unusually shallow expanse of water, with an average depth of about eighteen feet and a maximum depth of thirty-three, depending on the year. Though shallow, the lake is thirty-five miles wide and seventy-five miles long, making its surface area around one million acres during an average year. Because the lake resides on a relatively flat desert playa, these dimensions can change drastically. During high-precipitation years the lake can rise a few feet, which means the shoreline can extend by a few miles. In 1986/87, a historically high period for the Great Salt Lake, the area increased to almost 1.5 million acres as the shoreline stretched its fingertips into the surrounding land. During the historically low year of 1963, the area dropped to around 600,000 acres and the shoreline retreated dramatically, exposing miles of salty mud to the elements.

The Great Salt Lake is no stranger to massive fluctuations. Its predecessor was Lake Bonneville, a lake ten times larger than its size today and over nine hundred feet deep, deeper than the Great Lakes. Around fifteen thousand years ago a natural gateway dam broke in Red Rock Pass, Idaho. Water was released into the Snake River, Columbia Gorge, and eventually the Pacific Ocean at an estimated rate of fifteen million cubic feet per second. The Bible-worthy flood scoured the landscape for a few weeks after the dam broke. What was left of the lake was still much larger than it is today. After the last ice age the lake shrank again, until it reached its current size and shape.

Today's lake levels are a combination of natural geographical factors, the climate, and human influence. The average lake level is eleven feet lower today than it would have been if humans hadn't rerouted or dammed important tributaries. This makes the lake more susceptible to drought and further changes, since shallow bodies of water are more vulnerable to climate change than their deeper counterparts.

Other lakes around the world similar in depth and structure to the Great Salt Lake have already dried up due to climate change. The loss of the Aral Sea, once the fourth-largest lake in the world and located in the autonomous Karakalpakstan region of Central Asia, is perhaps one of the most devastating environmental losses of this century. Its name translates to "Sea of

Islands" because the inland sea once contained over one thousand islands. The Aral Sea is a sliver of what it once was, even with restoration efforts, and most of the human populations dependent on it have had to vacate the area. Reckless diversion practices by the Soviet Union and an arid and extreme climate are to blame. As the lake shrank it became saltier, harming both the water ecosystems and the surrounding ecosystems. The exposed dirt was bombarded by weapons testing, and the resulting dust storms made local agricultural land infertile and led to an overwhelming amount of health problems.

Owens Lake in Southern California went through a similar story. Greedy diverging practices drained the lake in the early 1900s and it was unable to recover naturally. The city of Los Angeles pumps twenty-five billion gallons of water into it every year to try to replenish Owens Valley, but their efforts are mostly futile. The dry lakebed is now the largest single source of dust in the United States, creating dust storms that harm agriculture, infrastructure, and human health. Over the past fifteen years Los Angeles has spent over $1.3 billion attempting to counteract the side effects of the dried lake.

The Great Salt Lake could meet a similar fate. Water diversions from Bear Lake and Bear River, the lake's largest tributary, have already lowered lake levels. Bear Lake is a consistent source of water for Northern Utah and Southern Idaho, filled each year by winter snowpack, but with low snow years, and river dams on the horizon, the main source of moisture for the Great Salt Lake and the ecosystems it feeds are at risk. If lake levels get too low and can't rebound, ecosystems dependent on the Great Salt Lake could collapse. Brine shrimp, which support the migrating bird populations, can survive in water with 25 percent salinity. The main basin in the lake currently fluctuates between 5 and 27 percent salinity. If the lake shrinks and becomes saltier, brine shrimp populations could suffer from the change in chemistry.

Rivers deposit over one million tons of minerals into the lake each year. When Utah was being mined in the 1800s, mine tailings were washed into the waterways and deposited on the shores of the Great Salt Lake. The saline lake bed contains centuries' worth of heavy minerals like mercury that could become toxic dust, harming residents of neighboring valleys and the snowpack. Lake effect storms would lack a lake to gather moisture from.

So much relies on the Great Salt Lake, yet it's largely ignored by residents of the surrounding metropolitan areas. It's often regarded as a nuisance due to the smells that can sometimes waft through the valley. Bird-watching and rowing are often the only forms of recreation centered around the lake. If Utahns want to fish, boat, float, or walk around a body of water in Northern Utah, they usually go to one of the local freshwater reservoirs, or they visit the shores of Utah Lake, a shallow, freshwater lake located in Utah Valley near Provo. When Utah Lake levels drop from diversion practices and drought, the resulting toxic algae blooms can kill animals, sicken humans, and poison crops.

I didn't know what Antelope Island was and didn't visit the Great Salt Lake until I was a teenager on a bird-watching field trip. Many Salt Lake City residents don't realize how large the Great Salt Lake is, or what a beautiful setting Antelope Island is for a herd of bison. Some residents of Ogden, the closest city to the lake, think the Great Salt Lake ends at Antelope Island, unaware that the majority of the lake extends for miles on the other side. Flying over the lake is the only way to truly witness the vastness of it. The next best way is to hike to Frary Peak on Antelope Island, where hikers are rewarded with 360-degree views.

Despite my desire to have a solitary hike to Frary Peak, there were a surprising number of families hiking that weekday. Seeing the herds of children alongside the herds of bison was heartwarming. I considered it an encouraging sign that so many people were visiting the stark landscape. Though being aware of and familiar with an area doesn't guarantee a willingness to fight for its survival, it seems that many people are more aware of threats to places they know than ones they are unfamiliar with.

I was pulled out of my reflective reverie by one of the families. A group of three or four kids ran down the trail behind me, but instead of passing they hovered at my heels, speaking a language that seemed to be either French or something they'd made up themselves. I couldn't tell which.

Irritated, I stepped to the side. As they passed me the first one, a young girl with long dark hair, said, "Merry Christmas!"

Before I could internally check myself and confirm that it was definitely *not* Christmas, a boy around the same age yelled "Happy Thanksgiving!" as he trotted by, followed closely by a "Happy Halloween!" from another dark-haired girl.

I couldn't tell whether I was annoyed or amused, but I let them get ahead of me. Within a few minutes they'd gone off-trail, and I was left no choice but to pass or awkwardly wait for them to continue. I didn't pull over when they caught up with me again. Every few minutes we met other groups of hikers traveling uphill, which always merited a "Merry Christmas!" "Happy Thanksgiving!" "Happy Halloween!" from the kids.

The entire time I wondered, *What did I do to deserve this?*

The answer came easily. When we were children, my sister and I would hike with our cousins from Provo and do the *exact* same thing, singing nursery rhymes at the top of our lungs. Other hikers must have despised us. After spending half a mile with someone else's kids tailing me, I was amazed at how tolerant my parents, grandparents, uncle, and aunt were during the fifteen-mile hikes they'd drag us on.

But of course, I knew why they let us sing. We were entertained; it made the hiking seem less of a chore. And they preferred us singing to us whining.

The shard of glass around the island, holding the water, earth, and sky within it, sparkled with a reflective intensity. By the end of the trail I was tempted to add "Happy Saint Patty's Day!" to their holiday greetings, but refrained. My throat had constricted, like it does whenever I think too hard about children in wild spaces, making connections with the land and with each other.

But that day, for some reason, the knot wasn't so bad. That morning, Colin and I had somehow ended up talking about parenting, about the different ways each of our parents went about raising us. A question arose in my mind. I was nervous to vocalize it, scared of how he might answer and scared to admit that I had been thinking about it. But if I didn't ask I knew it would haunt me the rest of the day.

"Do you think we'd be good parents?"

Colin didn't hesitate, so nonchalant with his answer that I knew he had thought about it before that moment.

"Of course we would."

A young climate activist and one of my friends, Brooke Larsen, wrote an essay published in *High Country News* titled "What Are We Fighting For?" In it she addresses one of the obstacles I feel when thinking about climate change: "Coming of age in an era of dystopian politics and climate chaos, it's easy for me to say what I'm against." Sometimes it feels like my adversaries, my worries, are too close. I focus on them rather than the things that give me hope. When thinking of climate change on too large a scale I can feel like I'm spinning in circles, losing all sense of self and place, becoming disoriented. It's when I can act like a ballerina, finding something in my vision that I can focus on and keep returning to while spinning, that I can choose hope over despair.

Declaring that I won't start a family because of climate change was never an easy thing to say, but it was much easier before I fell in love. Now, with someone I may want to share the rest of my life with, I don't know if I can keep to the promise I made that November night. The deep desire within me might outweigh my fear of risk, of the threats that will arise in my children's lives during the upcoming era of climate catastrophe. But isn't that what life is at its most basic level: bringing offspring into existence despite the perils of death and suffering?

Losing the snow will be much harder if I have children and grandchildren. Maybe that's how my whole decision to not have a family arose; I want to be able to share snow with my children and grandchildren like I was able to share it with my parents and grandparents. Maybe part of me is scared that I won't know how to raise them without snow to anchor our family to the Wasatch. But if I do decide to have children and the snow doesn't return, I'll find other ways of connecting them to this landscape I've fallen in love with. I'll do my best to enhance their world with love and light and laughter.

Migrating

A chorus of zipping noises cut through the air as we stripped out of our windbreakers.

"Anyone need sunscreen?" my mother asked, pulling a tube out of her backpack. Three hands reached out at once, and the tube began circling around our group, spitting globs of white onto palms before being slathered onto noses and cheeks.

We were at the top of Hidden Peak. My father was letting ski patrol know that our group was heading into the backcountry west of Twin Peaks. We had dropped a car off at a trailhead just down the canyon from Snowbird, where we'd be ending our day. Inside the car were camping chairs, cheese and smoked salmon appetizers, and a cooler filled with cold beer and champagne: our rewards for completing the hike we were about to embark on.

"Look, Pipeline's open," I said, and we all looked at the iconic chute across from us. Colin let out a low moan.

"Should we do that instead?"

Pipeline is one of the most challenging runs accessible from Snowbird. It requires a forty-five-minute hike along an exposed ridgeline to get to the top of the peak, and then a precarious route from there to the entrance. Depending on the season, it may open a few times during the spring, or it might not open at all. If it does open, it's only for an hour in the early morning.

Neither Colin, Tyndall, nor I had skied Pipeline. The last time it had opened Colin and I missed it by a half hour. Now, with it open, directly in front of us, our backcountry gear on our backs and beacons beeping in our pockets, the idea of finally skiing the chute was tantalizing.

When Snowbird opened in the 1970s, Baqui became the first person to ski Pipeline. The last time he skied it was in celebration of his eightieth birthday. My grandmother, father, uncle, and mother have all skied it. It's as though Pipeline is a rite of passage, a birth canal between two cliffs that could pop me out as a true Bounous. For Tyndall and me, it feels like a part of our legacy that we need to complete.

"You guys ready to go?" my dad asked as he sauntered up. While some people can look awkward walking in ski boots, my dad looks more natural and coordinated in them than he does in normal shoes.

"Pipeline's open," I said, unnecessarily.

"Yep," he said, "but they're closing it soon. We want to get more than just one run since it's our last day on skis."

We skied down a cat track to the entrance of the hike. It was the same gate that could take us to Pipeline, though our path would cut off in a different direction before we'd get to the summit of Twin Peaks. Strapping our skis onto our backpacks, we began kicking our toes into the steps patrol had packed down, climbing toward the sky. Conversations died as the air thinned. In minutes we were sweating through our layers. Some sections were so steep that if I bent my head forward a foot or so my nose could touch the snow.

It took about a half hour to reach the false summit, where our path split away from the trail that could have taken us to the top of Pipeline.

"Next year," Colin said wistfully.

The domed sky engulfed us. We were higher than most peaks in the area. We spent ten minutes appreciating the climb, the view, and two beers that Colin pulled out of his pockets before my dad urged us to step into our skis. The sun was powerful that spring day, and the snow was heating up quickly. We'd have to keep moving if we wanted to avoid avalanche danger.

Spring skiing can be fickle. When the days lengthen and the sun's strength returns, the top layer of the snowpack will partially melt in the heat of the day, then refreeze at night when the temperatures drop below freezing. Spring skiing can be dangerous in the morning for this reason, even at resorts, because the snowpack will be rock hard until it begins to melt. A simple fall could break a wrist, a leg, or even lead to a concussion or death. As the mountain warms, the snow degrades, turning the base of resorts into puddles, making the snow seem sticky, and creating wet avalanches.

The snow that we were aiming to ski is called *corn snow*, and is "Goldilocks" in character—not too hard, not too soft. When the snowpack undergoes multiple rounds of freezing and thawing it compresses, making it denser and increasing the water content. When the snow freezes overnight it creates that hard surface layer. During the day when the top of that crust begins to melt, water permeates through the crystals of ice, creating small balls called *large-grained* crystals. The consistency of these grains makes the snow soft and malleable, velvety and smooth to ski. Because the layers of snow underneath are still frozen, only the top few inches soften initially. As more snow crystals melt and water percolates through deeper layers, the snowpack becomes less stable and prone to "wet avalanches," which are slower moving and heavier than their dry counterparts.

Our early start was to avoid conditions that might lead to wet avalanches. So, after applying another layer of sunscreen, we pushed ourselves into motion, the morning sun casting our long shadows ahead of us, a few ski lengths in between each person.

At one point, my father, who was breaking trail, stopped. Our destination was beyond a ridge a hundred yards away. A steep, south-facing hillside stretched between us and the ridge. The reflection of the sun off the snow was blinding, as if a mirror lay on top of the pitch. We were sweating in our ski pants. The snow almost seemed to be steaming from the heat.

"Okay." My dad twisted his upper body around so we could hear him. "We'll take this hill one at a time. Wait until the person in front of you gets at least three-fourths of the way to the ridge before starting. Aim for the outcrop. And move quickly, the sooner we can get past this hill the better."

"Everyone's beeping?" asked my mom's voice from behind Colin.

"Yes," we all said, a tinge of exasperation in the response. My sister glanced at me, rolling her eyes with a smile. My mother had quadruple-checked that everyone had a working avalanche beacon before we had even boarded the tram.

My father pushed himself forward with his poles, moving his legs as if he were walking. As his skis cut tracks they displaced small nuggets of snow, which picked up speed and more snow as they rolled down the hill. Within a matter of seconds they'd build exponentially, until they weren't so much small balls as they were wheels of snow, some of them growing two feet in diameter before breaking apart with momentum.

Tyndall turned around again, her mouth pulled into a comical, clown-like frown.

"I guess this is where the term *snowballing* comes from."

She shuffled into motion and her long braid swung across her shoulders, shimmering gold in the sun. A cascade of snow wheels followed her skis. While some bounced down the pitch, others kept steady pressure on the hill. Their long, thin tracks looked like dark paths of meteorites across a white sky.

Then it was my turn. I tried not to focus on the shower of snowballs leaping from my skis. Instead I became intensely aware of how much snow was above me, how dense and heavy it was. If it all broke free at that moment . . . I tried to push the thought from my mind. My heart tapped at the base of my throat. It felt like the mountain was an immense, sleeping beast blocking our path, and we were trying to walk over its back without disturbing it. It was only when I stopped moving that I realized I had been holding my breath, as if breathing could have set off an avalanche.

The act of skiing is the act of taking a risk. It's a risk that we prepare for, and we do what we can to mitigate the hazards. But skiing is inherently dangerous. I can get annoyed when my mother nags about safety, but I share her worrisome genes. I hear the same exasperation in Colin's voice when he quips to my safety concerns that my own voice carries when responding to my mother.

In the past, people risked their lives on skis not for the pleasure of it, like we do, but because their work demanded it. Mail carriers used skis when the snow became too deep for horses or carriages, and many didn't survive. They were regarded as heroes in small, isolated mountain towns in the Colorado Rockies and Sierra Nevada, braving storms and passes to deliver mail, often alone in their endeavors. In an issue of *Colorado Magazine*, a quote from 1891 says that "usually there [was] a crowd at the post office to wish [the mail carrier] good luck."

Miners used skis to travel from their lodging to the mines and back. Avalanches posed a threat while they commuted, and were often still a threat when they returned home. Once mining claims were staked and mines excavated, the resulting towns built to accommodate the miners required lumber. Trees act as anchors for snow on steep slopes, as do boulders and outcrops, like the one my family aimed for that spring day. The more trees on a hill, the less likely the snow will slide and build enough momentum to become a proper avalanche. Usually when avalanches rip out trees it's because the avalanche starts above them.

When Alta was built, miners cut most of the old-growth forest that rooted the steep hillsides of the canyon. Without the trees to anchor the snow, massive avalanches would careen down the slopes, burying much of the town every winter. It's uncertain how many people died from avalanches during Alta's mining history. There are accounts of avalanches that killed 15 people at once, and almost 150 may have died between 1872 and 1911.

With a healthy forest and ski patrol who are experts in avalanche control, the risk of dying from an avalanche has greatly decreased. But warmer temperatures in the winter and rain, both side effects of climate change, could affect the stability of the snowpack. Avalanches may become more frequent and less predictable. When combined with population increase and more individuals venturing into the backcountry, the death toll from avalanches could increase in the future.

That is assuming that there will be enough snow in the backcountry to ski. While resorts will be able to rely on snowmaking to compensate for a decrease in natural snowfall, terrain outside of resorts is solely reliant on storms. More research has been conducted on how climate change will impact resorts than how it will impact the backcountry, but if resorts can expect seasons that are cut in half, we can expect similar impacts in backcountry areas.

We all got over the ridge safely. On the other side, a basin fanned out into long, rolling hills. We had miles of skiing before we had to take off our skis and hike up to the parking lot. Since this side of the mountain was north-facing, the snowpack wasn't as rotten as the last stretch had been. The snow made us feel like heroes, supporting our weight as we swished through the top few layers of soft crystals, chasing each other through the trees. It was wildly liberating, not following any designated runs. I darted between pines, trying to keep my family in sight, and felt as though we were a pack of wolves.

We didn't run into any other groups until we made it over the lip of the basin, the hanging valley of a glacier long gone, when we ran into the wide hiking path that would take us to the parking lot. There was still snow on the path, and we passed a group of schoolchildren, about eleven or twelve years in age, hiking up the trail, who paused throwing snowballs at each other just long enough for us to pass. Eventually the trail flattened out, we navigated through a couple of puddles in our skis, crossed a bridge over the swift waters of snowmelt, and hiked a short ways to the parking lot. Beer cans hissed as they were cracked, the pop of the champagne echoed around the aspen grove. We toasted to the end of our season.

After some appetizers, Colin stood up from his camping chair.

"Thanks for the adventure, I need to be heading south."

Tomorrow was Colin's first day on the river. He had packed his car with his river gear, leaving the mountains to head straight to the desert.

Something caught in my throat as I stood up to hug him goodbye. It was the end of another season together. The next few months would be bookmarked with stretches when we wouldn't see each other for days or weeks, another summer apart. This time, at least, our relationship had settled, compressing on itself like the snowpack through the winter. The last day of our ski season marked the start of a new season in our relationship.

My mind raced through thoughts and words, trying to find something to say in this moment of goodbye.

"Be safe," I said in his ear.

And then he was closing the car door and turning out of the parking lot, leaving us to our salmon and champagne.

Drowning

I always fight the arrival of spring. The appearance of buds along the gnarled branches of oaks makes many happy at the promise of warmer and longer days, but I become anxious when I realize my days skiing are numbered. I worry about when I'll next see a snowflake fall.

Depression that accompanies the passing of seasons is a medical condition called *seasonal affective disorder*. Winter is the most common time for this seasonal depression, when the days become shorter and colder, though the time of year can vary by person. A friend in Oregon becomes depressed heading into summer because of the persistent smoke that traps her and her family in their home. Summer used to be their favorite season, when they'd hike to a hillside to watch lightning storms in the distance. Now summer smoke is so terrible that the family is considering moving out of Oregon.

Even I can feel resistant to the transition into winter, but having an activity to look forward to counteracts this. Colin, too, feels lethargic as the days shorten, but the feeling was much worse when he lived in the Pacific Northwest, when skiing wasn't as much a staple for him. For us, skiing is the anchor that makes our transition into winter easier. Moving into spring reverses this stability for me, not because I dislike spring but because it pains me to watch winter dwindle. I reject the changes in the landscape. I want fewer blossoms and more snow storms. But the passing of seasons is as unstoppable as halting the spin of the earth. When I struggle with the seasonal procession, I struggle with the changes that inevitably come with the passage of time.

One spring I accidentally dove headfirst into a puddle on my high school's campus, attempting to save a volleyball from touching the ground. I don't remember much of the event, but I do remember that as I rose out of the dirty water my white polo stuck to my torso in indecent ways. In a picture my friend took of me after the dive, I'm proudly grinning, legs apart and hands on my hips like a superhero, with two dark smudges of dirt on my shirt where my breasts had rubbed the bottom of the dirty puddle while the rest of me just touched water.

My appearance in the bathroom mirror shocked me; though I had started developing two years prior, this smudged dirt on my chest seemed more like proof of my changing body than anything else had at the time. It wasn't the first time I had been embarrassed by my body, but it sticks out in my memory as being one of the most jarring realizations of my womanhood.

I resisted development. I was not ready for my body to change, not ready to have so many emotions I couldn't control. There were days when I couldn't stop crying and had no idea why. I didn't understand the changes. I became depressed. The future scared me, my body scared me, my emotions scared me.

It can be hard to comprehend and accept change. Sometimes it happens easily, and can even be a relief. But I am rarely one to readily accept change, especially if it involves risk. I want every season to be an everlasting season, want to remain in each stage of my life and never move to the next. Even change on a smaller scale ruffles my feathers. Construction aggravates me, I hate getting my hair cut, it pains me to watch a patch of sagebrush get ripped up and replaced with grass.

I had a pregnancy scare once. It happened during the spring, when the landscape was torn between buds and snowflakes. My emotional landscape was torn, too. I wasn't ready to move into that phase of life yet, wasn't ready for budding life to grow within me. My body resisted the change.

Like the many Americans who believe the climate is shifting but don't think it'll affect them, I resist climate change. Some understand that we can no longer stop it, and argue that instead of fighting against it we should put our efforts into adapting; that if we focus too much on trying to preserve the past and present we won't be able to adapt to the inevitable future. I understand this argument—adapt, migrate, or die. We can't migrate to another planet. If we focus our efforts on preparing communities for climate change, we might be able to save a lot of lives that would otherwise be lost to ill-preparedness. But that would mean accepting that climate change is nonreversible, unavoidable. To me, that means accepting that snow, at least the snow that I grew up with, that my family was shaped by, will become a thing of the past.

In the winter of 2017, the best ski season of my life so far, I convinced myself that those months might be some of my last truly amazing powder skiing. I dropped obligations whenever there was a storm, taking advantage of what I thought would be some of the last bottomless powder days of my life. I couldn't tell if I was overreacting or not, and when I expressed my fears I felt ridiculous. When the 2018 season came around, which started out worse than almost any other season on record, I felt justified for my powder frenzy. But as the season passed, the snow arrived. There may have not been as many deep days as the previous season, but there were still enough to enjoy. And the following season of 2019 was even more snow-filled than 2017.

No one knows what the following seasons will be like. Winters are unpredictable. The fluctuations between seasons are natural, and life has adapted to these changes. But no matter if it's a high snow year or a low snow year, skiing anchors my family to the mountain. I wonder how long that anchor will last, as snow turns to rain in the coming decades.

Despite the ski industry's impact on mountain ecosystems and our climate, I consider skiing an act of love toward snow. One of my favorite authors, a former river guide in southern Utah, Ellen Meloy, said, "I write a book about a river and cannot tell if it is a love story or an obituary or both." When I carve my name into the mountainside with my skis, I declare my love to the sky, weather, slope, gravity, snow. At times I worry that this love note is also a note of farewell.

Flowing

It's our first day on the Green River, on a stretch in eastern Utah known as "The Gates of Lodore." Technically we're in western Colorado, but we'll be in Utah within a day. I'm sitting in the front of a raft, staring at the tumultuous river before me as my heart flutters in my chest. I'm nervous. I wanted to sit in the back of the boat. I had already claimed my spot. But the other lady onboard doesn't want to ride in the front either, and she's seventy years old and I'm in my twenties.

Our rafts are tied to the shoreline above our first stretch of rapids and one of the most technical parts of the trip, "Disaster Falls," so named because John Wesley Powell and his crew lost a boat there in 1869. I know the name because almost exactly a year ago a woman lost her life to this very same rapid after her boat flipped. We had just scouted the rapid, hiking over a mile through deep sand in ninety degrees. From my vantage on shore, the waves looked enormous. Bigger than enormous: impossible. Rocks constricted the entrance of the rapid, a large boulder to the right and smaller rocks to the left, creating a narrow passage for water and boats to rip through. A sustained wave, looking powerful enough to smash a tree trunk into a million splinters, churned in the center of the flow. The current moved so fast it was dizzy to watch.

Colin stands up from his seat at the oars and says, "Come here."

I lean backward and he grabs the top of my life jacket, pulling up sharply, the motion that he'd have to make if I were to fall out of the boat in the upcoming rapids. Without saying another word he cinches the straps tighter and returns to checking that everything is tied down.

Colin is the trip leader, and spent this morning violently vomiting. From three a.m. until eight a.m., when we had to get in the vans for the three-hour ride to the drop-off, he couldn't leave the bathroom. He slept the entire drive, and has barely been able to drink any water. Being familiar with the energy he normally exhibits, I can tell he's weak. I wish I could say I'm confident in his skills as a guide, but I can't.

Finishing up his preparations, Colin releases the raft from the shore and we drift into the current. The rapids come quick and unforgiving. Frigid water from the deep snowpack of the Colorado Rockies leaps over the side of the boat, shocking in the desert heat. As we approach the main feature of the rapid I look from the river to Colin. He's turned the boat slightly to the side and is rowing backward, pulling us away from a boulder so we'll hit the sustained wave beside it. His face is still pale, his jaw clenched. His green eyes have a piercing focus. As we enter the wave, I'm left breathless at the sight of him.

It takes him days to recover fully from the sickness, but his fatigue makes it easier for him to sneak away from guide duties to lie in the tent with me and listen to the sounds of the river. Despite the looming crimson canyon walls, the sky feels like it's all around us. Rain sprinkles on our heads and shoulders one minute while sunlight flickers on the surface of the water around the corner. Our shoulders and noses color from the sun even as thunder clouds darken the canyon in patches.

It's late June and high water. The river around us is full and flush with the snowmelt of an incredible season. Our passion for skiing has melted with the snowpack into the waterways and the rivers, so we chase it down desert canyons with rafts and oars. Instead of skiing under the pines and firs of the mountainsides, we float between the tamarisk and willows of the Colorado Plateau. Rather than hearing the soft coo of a mourning dove, we turn our heads in awe at the cascading songs of canyon wrens. Instead of plodding through the snow and the mud in stiff ski boots, our toes dig their way into the hot, pink sand. These landscapes look and feel drastically different, but they are shaped by the same thing: snow.

Though much of the water from the Wasatch Front makes its way into the Great Salt Lake, most of the snow that falls in the West gets funneled into the great rivers of the desert. Here, riparian habitats are synced with the pulsing heartbeat of the American West—the veins from the mountains carry nutrients and water into this dry landscape and the vast artery continues carrying them through five US states before crossing the border into Mexico, depositing whatever's left into the Sea of Cortez.

As the snowpack melts, it fills the streams and creek beds, which then fill rivers and lakes. Vegetation along these pathways relies on an influx of moisture during the spring and early summer months, known as *runoff*. Spring floods cause soil erosion in certain areas of the floodplains and then redeposit this sediment elsewhere. This newly tilled soil, free from established plants, is fresh terrain for seedlings to sprout and grow rapidly. Native plants living along these pathways time their biologic clocks with the floods, releasing their seeds at the end of spring runoff when the river has calmed and newly made sandbars are shining and smooth. A thinner snowpack or snowpack layered with dust melts earlier, throwing this timing off. Without the reseeding of native plants, invasive plants like tamarisk are quick to move into these uninhabited areas, beating out native vegetation and changing the way sediment gathers on riverbanks.

Less snow in the mountains also means less groundwater to feed waterways even after the snowpack melts. Year-round waterways could become ephemeral, or short-lived, carrying water for a brief amount of time before disappearing. Some ephemeral waterways might dry up completely.

Snowpack provides almost 70 percent of the water in the American West. According to a recent study by Berkeley Lab, the water content in snowpack in the Sierra Nevada could drop almost 80 percent by 2100. In the low-snow year of 2018, 90 percent of snowpack monitors across the West were below average, some dropping as much as 70 percent of their average levels. The amount of water content lost from snowpack in the West could be equivalent to the amount of water held in Lake Mead, the reservoir that stores the most water in the United States.

The stars sparkle through the mesh fabric of our tent. Colin is snoring softly in my ear, his arm thrown across my stomach. A bird is flitting between branches nearby, the rustles just audible over the whispers of the river. There's a slight chill in the air, a welcoming coolness after the heat of the day.

I take a slow breath, bringing in the dry desert air through my nose, tasting the fine sand in the back of my throat. This is a new landscape for me, but with the warmth of Colin's body it feels familiar.

He follows water. Whether in the shape of the ocean tides ripping around rocky shorelines in the San Juan Islands, the cold smoke of the Wasatch Front, or as the sultry pulse of the Green and Colorado Rivers, Colin's true love is water, hydrogen bonds that make life on earth possible. Like his personality, a storm that doesn't settle on one landscape for too long before moving to the next, he exists not in one location but in a cycle. He follows the snow. His soul's place is not a place—it's the motion of life.

That night I have a dream I am melting. My body dissolves and a landscape rises around me. Where my skin flows, soil and wildflowers follow. My hair comes out in long strands, hardening into tree bark and pine needles. Leaves fall from my mouth as I exhale, scattering across roots that used to be toes. Tendons melt off my limbs like mud, exposing granite bones. My blood doesn't run in liquid form, but floats away from my body like snowflakes caught by a gust of wind.

Burying

It was the wing that caught my eye. Outstretched and upright, the tips of the feathers fanned out and up to the sky, poised and graceful even in death. It only took me a second to decide. I pulled the car over to the side of the road as I explained to my mother what I had seen.

"It's pretty big. I think it might be a hawk."

We stepped out of the car, grateful that no semi-trucks were behind us on this backcountry highway. The bird was only two feet from the edge of the pavement, its wing cocked at an unnatural angle. The way the body twisted on the ground made it look like a dancer, arms stretched in opposite directions, one facing up, one down. The underside of the wing facing us, the one raised to the sky, was a soft white, speckled with seeds of brown. Four tail feathers stuck out from beneath the bird, pointed like daggers. Its body was a myriad of golden hues. It looked as though it had fallen from the sun with broken wings. Icarus's wings.

I knelt on the ground beside it. There was dried blood where bent wing met body. Its white moon face was turned to the ground, one dark, almond-shaped eye visible. It was a barn owl, as silent in death as it would have been in life.

My mother and I drove to the nearest convenient store to buy garbage bags and rubber gloves. I slipped the gloves on my hands as a precaution, wishing I could feel the softness of its wings on my skin rather than the sticky rubber. As gently as I could, I slid my hands beneath the owl's wing bones, cradling its broken body. I nudged the wings inward as I lifted it off the ground and they folded easily, cocooning the bird's breast. Specks of down fell here and there upon my coat as I carried it to the car. It felt small in my arms, like an infant.

We drove up a canyon until we found a dirt road leading to a small, wooded area. The sun had already set behind the mountains and the hills around us were mauve with alpenglow. There was a frozen pond just off the road surrounded by leafless, deciduous trees. My mother wasn't wearing snow boots so I carried the owl to the base of one of those trees, placing it on the snow to rest while I dug a spot for its body. Against the white it looked like a piece of sunlight turned solid, its single, dark eye an entire universe. Once my fingers reached soil, I picked the owl up again and laid it in its grave. The shape of its body left an imprint in the snow.

"Do you want to say anything?" I asked my mother. She didn't hesitate.

"Who flies with beauty has no need of fear."

I buried it beneath the snow—the burial of the sky in the earth. I unearthed a few leaves as I dug and left them on the surface. They were the shape and color of the owl's feathers.

I turned back to my mother. The sky behind her was the color of marigolds. She smiled, kind wrinkles around her eyes as I walked back through the snow.

"Ready?" she asked. I nodded, and together we drove down the serpent curves of the canyon.

VI. Budding Season

Choosing

We step out of the car, our shoes padding softly onto the fresh inch of snow. I tuck the scarf ends into my coat and put my hands in my pocket, wishing I had chosen a slightly warmer jacket today. The weather of the last few weeks had been warm and wonderful. I thought that spring had arrived in full, but this last winter storm caught me and the changing landscape off guard.

The house in front of us is familiar, one I have spent hours looking at pictures of online. It looks small from the outside, a bungalow whose shingles have been painted dark blue and the accents around the door and windows a creamy beige.

We came without our realtor today. It's a spontaneous visit; we had been driving through the area and thought we should see the house one final time before we made a decision. We don't have the code to get inside so we stand in the backyard, looking at everything from the fruit trees, to the covered patio, to the garage doors, to the twisting, barren grapevine. The unobstructed face of Mount Olympus looms over the neighborhood, its deep wrinkles etched with white.

I've been considering what it would mean to commit to living in Utah, to *really* commit. I become nervous when thinking about the future of the Salt Lake Valley and the Wasatch Front.

The inversions during the winter and ozone pollution during the summer are already terrible, and our politicians are in a gridlock of complacency when it comes to improving air quality. It will certainly keep getting worse. If water keeps getting diverted from the Great Salt Lake tributaries, it may only be a matter of time before the flat expanse will dry up, leaving a salty playa behind. Dust storms will whip through the valley. And the snow. The snow . . . who knows what will happen with the snow? The future is filled with uncertainty.

"Well," Colin asks hesitantly, "what do you think. Should we do it?"

A veil of snowflakes has formed between us. It seems as though they just materialized out of the air. They are so small we can't see their shapes as they shimmer there, but I know their six points, those crystals that come down from heaven and look like stars.

I imagine the life that we could lead here. Late summer dinners out on the patio, picking peaches and raspberries from the garden bed behind the garage. How the leaves of the ash tree we stand under would scatter across the lawn when the canyon breezes come in the fall. I imagine lying in the master bedroom with him, watching the flakes of the first winter storm fall just outside the window, how the mountains we could see from bed would blush with alpenglow after a fresh snowfall. We'd have a long driveway to shovel when the storms really began rolling in, but a good garage to store our skis and boots in and an easy commute to the mountains on powder mornings. There are already crocuses and tulips staggered around the yard, adding pops of color against the inch of fresh snow.

Yes, I can imagine a future here with him, a life lived in the shadow of the Wasatch. But that's not all. There's a smaller, mint-colored bedroom on the first floor. When I walk into it I see a nursery, decorated with stars and snowflakes that hang from the ceiling, and a child with golden curls and eyes like the sky.

❅❆❅

A curious sound starts from a garden bed around the back of the garage, almost like a squeaky toy or bouncy ball. A burst of feathers startles both of us, and a California quail lands on a fence nearby. It's a male, the dangling black finger on its head shaking from its sudden flight. It eyes Colin and me suspiciously, tilting its head in jerky motions. More quails emerge from around the corner, pecking at the ground and making their little *tut-tut-tut!* noises.

"Oh," I say softly, "look, they have chicks."

A few balls of fluff are among the procession, so small they might have emerged from their eggs the day before. Adults keep to the edges of the group, creating a protective barrier around the chicks as they move. California quails are permanent residents of the Salt Lake Valley, no matter the season. They don't migrate and they don't hibernate, living through the winter in coveys of up to one hundred individuals. Multiple families live together at once, all parents caring for the chicks. They are territorial, dedicated creatures. Once they find a home that suits them, they settle in and can remain there their entire lives. And once they find a mate, they mate for life.

Chi-ca-go! Chi-ca-go! The male puffs out its blue chest and ruffles his chestnut wings. Its high-pitched call is one I know well, one I had even practiced in my youth so I could mimic the bird. I call back to it.

Chi-ca-go! Chi-ca-go!

Colin starts laughing, spooking the male back into flight. It flies across the yard, aiming for the fence on the far side. California quails have short, fat wings, good for lifting off the ground in bursts of energy to escape prey, but not good for flying longer distances. It doesn't quite clear the fence. The quail bumps into the wood panels and falls to the ground, immediately bursting into flight again. This time it makes it over the fence. Colin doubles over in laughter. Alarmed, the quails in the garden scamper around the corner and out of sight.

"Colin," I scold, though I'm laughing as well. "You scared them!"

He shakes his head, unable to speak for a few moments, eyes bright as he watches the quail procession scurry across the neighbors' driveway.

"What is it with you and little birds?" he says as we walk back down the driveway. "Maybe we can use them as bait to get some hawks to live in the backyard."

Of course the future is uncertain. No one can predict what will happen, even regarding climate change. The snow might disappear forever, or it may stay. Maybe my mother was right when she said that choosing to have children might be choosing to have hope. To bring life into this world is investing in the future, admitting that no matter the outcome it's still worth the risk. Author and activist Rebecca Solnit says that "to hope is to gamble. It's to bet on your futures, on your desires, on the possibility that an open heart and uncertainty is better than gloom and safety. To hope is dangerous, yet it is the opposite of fear, for to live is to risk."

I know what risk is. I've lived in the mountains my whole life. Risking is living, riding out the storms. And falling in love is to risk it all and commit your life to a cause, a landscape, another human, a family, knowing that one day you might lose something you love.

A few days later we decide to buy that house. Perhaps we will choose to have children one day, raise a family in this home and this landscape, or perhaps not. We will watch the family of California quails grow in size, their chicks turning from tiny things of feathers in the spring to adults by the end of summer. They visit our yard frequently, in groups of thirty or more, pecking at the ground. Every time I hear *tut-tut-tut!* I rush to the window to watch them scatter through our yard. Colin stands next to me, laughing at how they seem to peck at nothing, how the males puff their chests out, looking so self-important, or when one of them hits the fence mid-flight. It happens quite often. Sometimes I'll be caught off guard in the garden when they emerge from behind the rose bushes, and I'll catch a glimpse of something profound in their eyes before they fly away in a burst of feathers.

It's the time we spend investing in the things that give us joy, the things that make us who we are, that will continue to give us the strength and courage that we need to keep fighting. Be that a landscape, a relationship, a community.

Either way, whether or not I choose to become a mother, I will try to choose hope from now on, rather than despair. I will choose starlight, aspen leaves, mountain peaks, snowflakes.

ACKNOWLEDGEMENTS

I've wanted to be an author ever since I was ten years old, when I listened to my cousin Evalynn read a story she had written while we were camping in the Needles District of southern Utah. It's taken a village to help me write my first book, and quite a few thanks are in order.

So thank you . . .

To those who have taught me, both in and out of the classroom, that writers can be naturalists and that words can be our weapons against climate change, including but not limited to: Juanito, Terry & Brooke Williams, Jeff McCarthy, Sylvia Torti, Brett Clark, Steve Tatum, Ben Reeder, Ellen Meloy, Steve Trimble, Michael Mejia, Rebecca Solnit, Dolores LaChapelle, and of course my EH peers—the greatest teachers I could ask for during these uncertain times.

An extra thank you to Jeff, for believing in my potential right from the beginning and, upon hearing my early ramblings for this story, for encouraging me to pursue a master's project dedicated to snow.

To the women at Torrey House Press, for their patience, kindness, passion, and drive. I'm proud to call myself a Torrey House author because of you.

To Snowbird Ski Patrol, for keeping my family, friends, and me safe on our favorite days of the year.

To the Gaylord family, for welcoming me in and lending me snippets of their family story. May there always be a Bounous and a Gaylord together at Snowbird.

To the Anderson family, for their love and continued support as a second family, and especially to Shelley and Whitney, for gifting me a journal and therefore the gift of storytelling for every birthday.

To my cousins—all of them—for our adventures.

To Colin, for letting me write so much about his hair.

To my sister, for being strong, brave, and my best friend.

To Grant and Suzanne, for teaching me how to love and how to sing.

To Junior and Maxine, for teaching me to love wildflowers and powder skiing.

To my parents, for their unwavering support and unconditional love. I would not have been able to fulfill this dream without them.

SOURCES

"A Short History of Skis." *International Skiing History Association*, https://www.skiinghistory.org/history/short-history-skis-0.

Abbey, Edward. *Desert Solitaire*. Tucson: University of Arizona Press, 1988.

"Air Pollution and Public Health in Utah." Utah Department of Health, https://health.utah.gov/utahair/pollutants/.

Anderegg, William R. L., et al. "Meta-Analysis Reveals that Hydraulic Traits Explain Cross-Species Patterns of Drought-Induced Tree Mortality across the Globe." *Proceedings of the National Academy of Sciences* 113, no. 18 (2016): 5024–29. doi:10.1073/pnas.1525678113.

Anderegg, William R. L., et al. "The Roles of Hydraulic and Carbon Stress in a Widespread Climate-Induced Forest Die-Off." *Proceedings of the National Academy of Sciences* 109, no. 1 (2011): 233–37. doi:10.1073/pnas.1107891109.

"All About Snow: Snow Characterstics." National Snow and Ice Data Center. https://nsidc.org/cryosphere/snow/science/characteristics.html.

"Alta Historical Timeline." AltaCam, https://www.altacam.com/local/alta-history.html.

Arndt, Deke. "Alaska: Last Frontier on the Front Lines." Climate.gov, NOAA, 20 May 2016. https://www.climate.gov/news-features/blogs/beyond-data/alaska-last-frontier-front-lines-climate-change.

Ault, Toby R., et al. "Relative Impacts of Mitigation, Temperature, and Precipitation on 21st-Century Megadrought Risk in the American Southwest." *Science Advances* 2, no. 10 (2016). doi:10.1126/sciadv.1600873.

Barnett, T. P., et al. "Potential Impacts of a Warming Climate on Water Availability in Snow-Dominated Regions." *Nature* 438, no. 7066 (2005): 303–09. doi:10.1038/nature04141.

Beever, Erik A., et al. "Pika (*Ochotona Princeps*) Losses from Two Isolated Regions Reflect Temperature and Water Balance, but Reflect Habitat Area in a Mainland Region." *Journal of Mammalogy* 97, no. 6 (2016): 1495–1511. doi:10.1093/jmammal/gyw128.

Bentley, W. A., and W. J. Humphreys. *Snow Crystals*. York: Maple, 1931.

Berwyn, Bob. "Polar Vortex: How the Jet Stream and Climate Change Bring on Cold Snaps." InsideClimate News, 2 Feb. 2018. https://insideclimatenews.org/news/02022018 /cold-weather-polar-vortex-jet-stream-explained-global -warming-arctic-ice-climate-change

Branch, John. "The Mush in the Iditarod May Soon Be Melted Snow." *The New York Times*, 1 Mar. 2019. https://www .nytimes.com/2019/03/01/sports/iditarod-climate-change -warming.html.

Brown, Ross D., and Philip W. Mote. "The Response of Northern Hemisphere Snow Cover to a Changing Climate." *Journal of Climate* 22, no. 8 (2009): 2124–45. doi:10.1175/2008jcli2665.1.

Burakowski, Elizabeth, and Matthew Magnusson. *Climate Impacts on the Winter Tourism Economy in the United States*. NRDC and POW, 2012. https://www.nrdc.org/sites /default/files/climate-impacts-winter-tourism-report.pdf

Byrne, Kevin. "Why Does It Become So Quiet after a Fresh Snowfall?" AccuWeather, www.accuweather.com/en /weather-news/why-does-it-become-so-quiet-after-a-fresh -snowfall/70000676.

Carling, Gregory Todd. *Trace Element Cycling in Great Salt Lake Wetlands, Wasatch Snowpack, and Mining-impacted*

Streams of Southern Ecuador. Salt Lake City: Department of Geology and Geophysics, University of Utah, 2012.

Cheshire, Laura. "From Shovels to Plows: A History of Snow Removal." Jalopnik, 18 June 2013. jalopnik.com/from -shovels-to-plows-a-history-of-snow-removal-5719723.

Coleman, Annie Gilbert. "From Snow Bunnies to Shred Betties: Gender, Consumption, and the Skiing Landscape." *Seeing Nature through Gender*. Lawrence: University of Kansas, 2003.

Coleman, Annie Gilbert. "The Unbearable Whiteness of Skiing." *Pacific Historical Review* 65, no. 4 (1996): 583–614.

Cook, Benjamin. "Guest Post: Climate Change Is Already Making Droughts Worse." Carbon Brief, 15 May 2018, https://www.carbonbrief.org/guest-post-climate-change-is -already-making-droughts-worse.

Davis, Robert E., Kelly Elder, Daniel Howlett, and Eddy Bouzaglou. "Relating Storm and Weather Factors to Dry Slab Avalanche Activity at Alta, Utah, and Mammoth Mountain, California, Using Classification and Regression Trees." *ScienceDirect*. Cold Regions Science and Technology, 1999.

Deems, J. S., et al. "Combined Impacts of Current and Future Dust Deposition and Regional Warming on Colorado River Basin Snow Dynamics and Hydrology." *Hydrology and Earth System Sciences* 17, no. 11 (2013): 4401–13. doi:10.5194/hess-17-4401-2013.

Denning, Andrew. "How Skiing Went from the Alps to the Masses." *The Atlantic*, Atlantic Media Company, 23 Feb. 2015. https://www.theatlantic.com/business/ archive/2015/02/how-skiing-went-from-the-alps-to-the -masses/385691/.

Dunfee, Ryan. "The Ski Industry Lobby for Climate Change." *POWDER Magazine*, 30 Oct. 2012. https://powder.com/

stories/climate-change-politics/#S3CcEH6JDF1D3DF7
.97.

Elliot, Kylke. "What Are High Pressure Systems and How Do
They Contribute to Our Weather?" Accuweather. https://
www.accuweather.com/en/weather-news/what-are-high
-pressure-systems-and-how-do-they-contribute-to-our
-weather/70005291.

England, Katie. "Snowbird Finishes up Mary Ellen Gulch
Projects Hoped to Improve Water Quality." *Daily Herald*, 26
Sept. 2018. https://www.heraldextra.com/news/local/north
/snowbird-finishes-up-mary-ellen-gulch-projects-hoped
-to-improve/article_b596ae3d-50a9-5f3b-afd8
-e8ebf9dbb42b.html.

"Fastest Avalanche." Guinness World Records, https://www
.guinnessworldrecords.com/world-records/fastest
-avalanche.

Flynn, Casey. "Cost of Snowmaking." ESPN, ESPN Internet
Ventures, 5 Jan. 2013, https://www.espn.com/action
/freeskiing/story/_/id/8809682/cost-snowmaking.

Fountain, Andrew G., et al. "The Disappearing Cryosphere:
Impacts and Ecosystem Responses to Rapid Cryosphere
Loss." *BioScience* 62, no. 4 (2012): 405–415. doi:10.1525
/bio.2012.62.4.11.

Fox, Porter. *Deep: The Story of Skiing and the Future of Snow*.
Jackson Hole: Rink House Productions, 2013.

Fox, Porter. "Ski Industry Leaders Linked to Climate Change
Deniers" *POWDER Magazine*, 3 Nov. 2016. https://www.
powder.com/stories/news/campaign-donations
-link-ski-industry-leaders-climate-change-deniers/.

Galatas, Eric. "Wildlife Biologist: Climate Change Threatens
Mountain Goats." Public News Service, 1 Apr. 2015. https://
www.publicnewsservice.org/2015-04-01/endangered
-species-and-wildlife/

Gheorghiu, Dragoș, and George Nash. *Place as Material*

Culture: Objects, Geographies and the Construction of Time.
Cambridge: Cambridge Scholars, 2013.

Gilaberte-Búrdalo, M., et al. "Impacts of Climate Change on Ski
Industry." *Environmental Science & Policy* 44 (Aug. 2014):
51–61. doi:10.1016/j.envsci.2014.07.003.

Glenn Albrecht Murdoch University. "The Age of Solastalgia."
The Conversation, 27 Feb. 2017. https://theconversation
.com/the-age-of-solastalgia-8337.

"Great Salt Lake, Utah." *USGS Utah Water Science Center: Great
Salt Lake.* https://ut.water.usgs.gov/greatsaltlake/.

Hatchett, Benjamin J., and Daniel J. McEvoy. "Exploring the
Origins of Snow Drought in the Northern Sierra Nevada,
California." *Earth Interactions* 22, no. 2 (2018): 1–13.
doi:10.1175/ei-d-17-0027.1.

Hausfather, Zeke. "CO_2 as a Feedback and Forcing in the
Climate System." Yale Climate Connections, 25 Oct. 2007.
https://www.yaleclimateconnections.org/2007/10/common
-climate-misconceptions-co2-as-a-feedback-and-forcing
-in-the-climate-system/.

Heidegger, Martin. *Being and Time.* N.p.: n.p., 1967.

"History." *Great Salt Lake Ecosystem Program*, https://wildlife
.utah.gov/gsl/history/index.php.

"How Aspens Grow." US Forest Service, https://www.fs.fed.us
/wildflowers/beauty/aspen/grow.shtml.

Howe, Peter D., et al. "Geographic Variation in Opinions
on Climate Change at State and Local Scales in the
USA." *Nature Climate Change* 5, no. 6 (2016): 596–603.
doi:10.1038/nclimate2583.

Hudson, Simon. *Sport and Adventure Tourism.* New York:
Haworth Hospitality, 2003.

Hudson, Simon. "The 'Greening' of Ski Resorts: A Necessity
for Sustainable Tourism, or a Marketing Opportunity for
Skiing Communities?" *Journal of Vacation Marketing* 2, no.
2 (1996): 176–85.

Hughes, Ryan. "Man-Made Snow: What Are the Environmental Effects?" The Inertia, 24 June 2016. https://www.theinertia .com/mountain/man-made-snow-what-are-the -environmental-effects/#modal-close.

Impacts of Climate Change on Water and Ecosystems in the Upper Colorado River Basin. USGS, 2007. https://action .suwa.org/site/DocServer/USGS_CO_Plateau_Climate .pdf?docID=4101.

Jones, Leigh Pender. "Assessing the Sensitivity of Wasatch Snowfall to Temperature Variaions." Thesis. University of Utah, 2010.

Kant, Immanuel, and James Creed Meredith. *The Critique of Judgement.* Oxford: Clarendon Press, 2010.

Kaufmann, Walter Arnold, and Martin Buber. *I and Thou.* N.p.: n.p., 1970.

Kepler, Johannes. *The Six-Cornered Snowflake.* Oxford: Clarendon Press, 1966.

Kimmerer, Robin Wall. *Braiding Sweetgrass: Indigenous Wisdom, Scientific Knowledge, and the Teachings of Plants.* Minneapolis: Milkweed Editions, 2013.

LaChapelle, Dolores. *Deep Powder Snow: 40 Years of Ecstatic Skiing, Avalanches, and Earth Wisdom.* 1st ed. Durango, CO: Kivakí Press, 1993.

LaChapelle, Dolores. *Sacred Land, Sacred Sex: Rapture of the Deep: Concerning Deep Ecology and Celebrating Life.* Durango, CO: Kivakí Press, 1988.

Larsen, Brooke. "What Are We Fighting For?" *High Country News,* 13 Nov. 2017. https://www.hcn.org/issues/49.19 /activism-what-are-we-fighting-for.

Larsen, Leia. "We're Losing the Great Salt Lake; Here's Why You Should Care." *Standard Examiner,* 11 Oct. 2015. https:// www.standard.net/news/environment/we-re-losing-the -great-salt-lake-here-s-why/article_0371d47b-0ac5-56dd -92ed-17eead22733c.html.

Larsen, Leia. "As Great Salt Lake Dries Up, Utah Air Quality Concerns Blow In." *Standard Examiner*, 11 Oct. 2015. https://www.standard.net/news/environment/as-great-salt -lake-dries-up-utah-air-quality-concerns/article_104452df -6fd1-5f2a-8601-0432cc1f5ad5.html.

Larsen, Leia. "Great Salt Lake Dust Model Spells Trouble for Utah If More Water Is Diverted." *Standard Examiner*, 29 May 2017. https://www.standard.net/news/environment /great-salt-lake-dust-model-spells-trouble-for-utah-if /article_ead11965-c5c1-5a64-a22f-7fdf1e351319.html.

Leopold, Aldo. *A Sand County Almanac and Sketches Here and There*. Oxford: Oxford University Press, 1968.

Libbrecht, Kenneth. *Field Guide to Snowflakes: Identifying Crystal Types, the Science behind Snowflakes, Observation Tools and Tips, a Close-Up Look at Nature's Art*. Minneapolis: Voyageur Press, 2016.

Lindsey, Rebecca. "Warming and Extreme Dust Could Advance Spring Thaw in Colorado Basin by 6 Weeks." Climate.gov, NOAA, 29 Nov. 2013. https://www.climate.gov/news -features/featured-images/warming-and-extreme-dust -could-advance-spring-thaw-colorado-basin-6.

Lindsey, Rebecca. "Climate Conditions behind Deadly October 2017 Wildfires in California." Climate.gov, NOAA, 11 Oct. 2017. https://www.climate.gov/news-features/event-tracker /climate-conditions-behind-deadly-october-2017 -wildfires-california.

Lohan, Tara. "How Climate Change Is Impacting the American West Right Now." News Deeply, 30 Nov. 2017. https://www .newsdeeply.com/water/articles/2017/12/04/how-climate -change-is-impacting-the-american-west-right-now.

McCarthy, Jeffrey Mathes. *Green Modernism: Nature and the English Novel 1900 to 1930*. N.p.: Palgrave MacMillan, 2014.

McCusker, Kelly, and Hannah Hess. "America's Shrinking Ski Season." Climate Impact Lab, 8 Feb. 2018. https://www

.impactlab.org/news-insights/americas-shrinking-ski
-season/.

Meloy, Ellen. *Eating Stone: Imagination and the Loss of the Wild.*
New York: Vintage, 2006.

Meloy, Ellen. *The Anthropology of Turquoise: Reflections on
Desert, Sea, Stone, and Sky.* New York: Vintage, 2003.

Meyer, Robinson. "The Southwest May Be Deep into a Climate-
Changed Mega-Drought." *The Atlantic*, Atlantic Media
Company, 18 Dec. 2018. https://www.theatlantic.com
/science/archive/2018/12/us-southwest-already-mega
-drought.

Monroe, Rob. "Note on Reaching the Annual Low Point." The
Keeling Curve, Scripps Institution of Oceanography, 23
Sept. 2016, https://scripps.ucsd.edu/programs
/keelingcurve/2016/09/23/note-on-reaching-the-annual
-low-point/.

Moore, Kathleen Dean. *The Pine Island Paradox.* Minneapolis:
Milkweed Editions, 2005.

Mote, Philip W., et al. "Dramatic Declines in Snowpack in the
Western US." *NPJ Climate and Atmospheric* 1, no. 2 (2018).
https://www.nature.com/articles/s41612-018-0012-1.

Mountain Research Initiative EDW Working Group.
"Elevation-Dependent Warming in Mountain Regions of
the World." *Nature Climate Change* 5, no. 5 (2015): 424–30.
doi:10.1038/nclimate2563.

Muir, John. *My First Summer in the Sierra.* New York: Modern
Library, 2003.

Murtaugh, Paul A., and Michael G. Schlax. "Reproduction and
the Carbon Legacies of Individuals." *Global Environmental
Change* 19, no. 1 (2009): 14–20. doi:10.1016/j
.gloenvcha.2008.10.007.

"National Ski Areas Assn Contributions to Federal Candidates,
2012 Cycle." OpenSecrets.org, https://www.opensecrets.org
/pacs/pacgot.php?cmte=C00327130&cycle=2012.

"SOTC: Northern Hemisphere Snow." National Snow and Ice Data Center. Feb. 2017. https://nsidc.org/cryosphere/sotc/snow_extent.html

Naff, Clay F. "Return of the Dust Bowl." *EARTH Magazine*, 27 Aug. 2012, https://www.earthmagazine.org/article/return-dust-bowl.

Nims, John Frederick. *The Six-Cornered Snowflake and Other Poems*. New York: New Directions, 1990.

Nixon, Rob. *Slow Violence and the Environmentalism of the Poor*. Cambridge: Harvard University Press, 2013.

"Northern Hemisphere Snow." National Snow and Ice Data Center, https://nsidc.org/cryosphere/sotc/snow_extent.html.

Northon, Karen. "Study: Carbon Emissions Could Increase Risk of U.S. Megadroughts." NASA, 19 Mar. 2015, https://www.nasa.gov/press/2015/february/nasa-study-finds-carbon-emissions-could-dramatically-increase-risk-of-us.

Nuccitelli, Dana. "The Many Ways Climate Change Worsens California Wildfires." Yale Climate Connections, 14 Nov. 2018, https://www.yaleclimateconnections.org/2018/11/the-many-ways-climate-change-worsens-california-wildfires/.

"Over and Alps: Can Artificial Snow Save Skiing in Europe?" World Travel Guide, 2 Mar. 2015. https://www.worldtravelguide.net/features/feature/over-and-alps-can-artificial-snow-save-skiing-in-europe/.

Page, David. "The Quiet Force." *POWDER Magazine*, 24 Mar. 2016. https://www.powder.com/the-quiet-force/.

Painter, Thomas H., et al. "Variation in Rising Limb of Colorado River Snowmelt Runoff Hydrograph Controlled by Dust Radiative Forcing in Snow." *Geophysical Research Letters* 45, no. 2 (2018): 797–808. doi:10.1002/2017gl075826.

"Pando." Western Aspen Alliance, Utah State University, https://western-aspen-alliance.org/pando/index.

"Particulate Matter (PM) Pollution." Environmental Protection Agency, 12 Nov. 2018, https://www.epa.gov/pm-pollution.

Patten, Duncan T. "Riparian Ecosytems of Semi-Arid North America: Diversity and Human Impacts." *Wetlands* 18, no. 4 (1998): 498–512. doi:10.1007/bf03161668.

Pierre-Louis, Kendra, and Nadja Popovich. "Of 21 Winter Olympic Cities, Many May Soon Be Too Warm to Host the Games." *The New York Times*, 11 Jan. 2018. https://www.nytimes.com/interactive/2018/01/11/climate/winter-olympics-global-warming.html.

Pleij, Herman. "Urban Elites in Search of a Culture: The Brussels Snow Festival of 1511." *New Literary History* 21, no. 3 (1990): 629–47.

Rice, Janine, et al. *Assessment of Watershed Vulnerability to Climate Change for the Uinta-Wasatch-Cache and Ashley National Forests, Utah.* USDA, 2017. https://www.fs.fed.us/rm/pubs_series/rmrs/gtr/rmrs_gtr362.pdf.

Rhoades, Alan M., et al. "The Changing Character of the California Sierra Nevada as a Natural Reservoir." *Geophysical Research Letters* 45, no. 23 (2018). doi:10.1029/2018gl080308.

Saeland, Jodi. "The Watershed underneath the Wasatch Mountains." KSL.com, 14 Dec. 2013. https://ksl.com/article/28033253.

Sanchez, Hayley. "Map: Income Inequality in Colorado Is Most Extreme Near Ski Towns." Colorado Public Radio, 26 July 2018. https://www.cpr.org/news/story/colorado-income-inequality-ski-towns.

Schaefer, Kevin. "Methane and Frozen Ground." National Snow & Ice Data Center, nsidc.org/cryosphere/frozenground/methane.html.

Scott, D., et al. "The Future of the Olympic Winter Games in an Era of Climate Change." *Current Issues in Tourism* 18, no. 10 (2014): 913–30. doi:10.1080/13683500.2014.887664.

Scott, Michon. "Climate & Skiing." Climate.gov, NOAA, 19 Nov. 2018, https://www.climate.gov/news-features/climate -and/climate-skiing.

Scott, Michon. "Mountain Air Becoming Less Brisk, More High Elevation Observations Needed." Climate.gov, NOAA, 23 Apr. 2015. https://www.climate.gov/news-features/featured -images/mountain-air-becoming-less-brisk-more-high -elevation-observations.

Skiles, S. McKenzie, et al. "Implications of a Shrinking Great Salt Lake for Dust on Snow Deposition in the Wasatch Mountains, UT, as Informed by a Source to Sink Case Study from the 13–14 April 2017 Dust Event." *Environmental Research Letters* 13, no. 12 (2018): 124031. doi:10.1088/1748-9326/aaefd8.

"Snowmelt." The Water Cycle, USGS, https://water.usgs.gov /edu/watercyclesnowmelt.html.

Solnit, Rebecca. *Hope in the Dark: Untold History of People Power*. Melbourne: Text Publishing, 2005.

Solnit, Rebecca. *Wanderlust: A History of Walking*. London: Granta Books, 2014.

Steenburgh, W. James. "One Hundred Inches in One Hundred Hours: Evolution of a Wasatch Mountain Winter Storm Cycle." *Weather and Forecasting* 18, no. 6 (2003): 1018–36. doi:10.1175/1520-0434(2003)018<1018:ohiioh>2.0.co;2.

Steenburgh, Jim. *Secrets of the Greatest Snow on Earth: Weather, Climate Change, and Finding Deep Powder in Utah's Wasatch Mountains and around the World*. Logan: Utah State University Press, 2014.

Stratus Consulting, Preparer. *Climate Change in Park City: An Assessment of Climate, Snowpack, and Economic Impacts*. 2009.

Stoddart, Mark C. J. *Making Meaning out of Mountains: The Political Ecology of Skiing*. Vancouver: UBC, 2012.

Sunyer, Jordi, et al. "Association between Traffic-Related Air Pollution in Schools and Cognitive Development in

Primary School Children: A Prospective Cohort Study." *PLOS Medicine* 12, no. 3 (2015). doi:10.1371/journal .pmed.1001792.

"Surface Hoar." Avalanche.org, https://avalanche.org/avalanche -encyclopedia/surface-hoar/.

Turville-Petre, E. O. G. *Myth and Religion of the North: The Religion of Ancient Scandinavia*. Westport, CT: Greenwood, 1977.

Weissbecker, Inka. *Climate Change and Human Well-Being: Global Challenges and Opportunities*. New York: Springer, 2011.

"Wet Snow Avalanche." Avalanche.org, https://avalanche.org /avalanche-encyclopedia/wet-snow-avalanche/.

Weuve, Jennifer, et al. "Exposure to Particulate Air Pollution and Cognitive Decline in Older Women." *Archives of Internal Medicine* 172, no. 3 (2011): 219–27. doi:10.1001 /archinternmed.2011.683.

Wilker, E. H., et al. "Long-Term Exposure to Fine Particulate Matter, Residential Proximity to Major Roads and Measures of Brain Structure." *Stroke* 46, no. 5 (2015): 1161–66. doi:10.1161/strokeaha.114.008348.

Williams, A. Park, et al. "Temperature as a Potent Driver of Regional Forest Drought Stress and Tree Mortality." *Nature Climate Change* 3, no. 3 (2012): 292–97. doi:10.1038 /nclimate1693.

Wobus, Cameron, et al. "Projected Climate Change Impacts on Skiing and Snowmobiling: A Case Study of the United States." *Global Environmental Change* 45 (2017): 1–14. doi:10.1016/j.gloenvcha.2017.04.006.

Yuhas, Daisy. "Storm Scents: It's True, You Can Smell Oncoming Summer Rain." *Scientific American*, 18 July 2012. https://www.scientificamerican.com/article/storm -scents-smell-rain/.

Zhang, Hongchao, and Jordan Smith. "Weather and Air Quality Drive the Winter Use of Utah's Big and Little Cottonwood Canyons." *Sustainability* 10, no. 10 (2018): 3582. doi:10.3390/su10103582.

Zinman, Jonathan, and Eric W. Zitzewitz. "Snowed: Deceptive Advertising by Ski Resorts." *SSRN*, 29 June 2009. https://papers.ssrn.com/sol3/papers.cfm?abstract_id=1427490.

ABOUT THE AUTHOR

Ayja Bounous is a Utah native and avid skier. She holds an MA in Environmental Humanities from the University of Utah and bachelor's degrees in Music and Environmental Studies from Santa Clara University. She lives in Salt Lake City, Utah.

ABOUT THE COVER

"The skier in this photo is Sammo Cohen, one of my longtime ski partners and friends. We shot this photo during an early morning ski touring outing in the Wasatch Mountains of Utah. We saw this shield of snow from where we were hiking, and it looked too good to pass by. Sammo hiked to the top, and laid down a set of perfect turns."

Sam Watson is a photographer and outdoorsman from Salt Lake City, Utah. A lifelong skier, biker, backpacker, hiker, and explorer, Sam specializes in shooting photos of the things he loves to do. With or without a camera, Sam can be found in the mountains or deserts of Utah and beyond.

Find more of Watson's work at www.samwatsonphotos.com.

TORREY HOUSE PRESS

Voices for the Land

The economy is a wholly owned subsidiary of the environment, not the other way around.
—Senator Gaylord Nelson, founder of Earth Day

Torrey House Press is an independent nonprofit publisher promoting environmental conservation through literature. We believe that culture is changed through conversation and that lively, contemporary literature is the cutting edge of social change. We strive to identify exceptional writers, nurture their work, and engage the widest possible audience; to publish diverse voices with transformative stories that illuminate important facets of our ever-changing planet; to develop literary resources for the conservation movement, educating and entertaining readers, inspiring action.

Visit www.torreyhouse.org for reading group discussion guides, author interviews, and more.

As a 501(c)(3) nonprofit publisher, our work is made possible by generous donations from readers like you. Join the Torrey House Press family and give today at www.torreyhouse.org/give.

This book was made possible by generous gifts from The Nature Conservancy of Utah, HEAL Utah, Tara Tull, Kristy Larsen, Judith Freeman, William Graney, Erika Eve Plummer, Marion Lennberg, and Don Gomes and Annie Holt. Torrey House Press is supported by Back of Beyond Books, the King's English Bookshop, Jeff and Heather Adams, the Jeffrey S. and Helen H. Cardon Foundation, the Grant B. Culley Jr. Foundation, Jerome Cooney and Laura Storjohann, Heidi Dexter and David Gens, Kirtly Parker Jones, Suzanne Bounous, Diana Allison, the Utah Division of Arts & Museums, and Salt Lake County Zoo, Arts & Parks. Our thanks to individual donors, subscribers, and the Torrey House Press board of directors for their valued support.

Join the Torrey House Press family and give today at www.torreyhouse.org/give.